Game Theory and Economic Modelling

Game Theory
and
Economic Modelling

DAVID M. KREPS

CLARENDON PRESS · OXFORD

Oxford University Press, Walton Street, Oxford OX2 6DP

Oxford New York Toronto
Delhi Bombay Calcutta Madras Karachi
Petaling Jaya Singapore Hong Kong Tokyo
Nairobi Dar es Salaam Cape Town
Melbourne Auckland
and associated companies in
Berlin Ibadan

Oxford is a trade mark of Oxford University Press

Published in the United States
by Oxford University Press, New York

© *David M. Kreps 1990*

First published 1990
Reprinted 1991 (twice)

British Library Cataloguing in Publication Data
Data available
ISBN 0–19–828357–1 (hbk)
ISBN 0–19–828381–4 (pbk)

Library of Congress Cataloging in Publication Data
Data available
ISBN 0–19–828357–1 (hbk)
ISBN 0–19–828381–4 (pbk)

Printed and bound in Great Britain by
Biddles Ltd,
Guildford and King's Lynn

Acknowledgements

This book provides a lengthened account of a series of lectures given at the Tel Aviv University under the aegis of the Sackler Institute for Advanced Studies and the Department of Economics in January 1990 and then at Oxford University in the Clarendon Lecture Series while I was a guest of St Catherine's College in February 1990. (Chapter 3 contains material not given in the lecture series that is useful for readers not conversant with the basic notions of non-cooperative game theory.) I am grateful to both sets of institutions for their hospitality in the event and for their financial support. I am also pleased to acknowledge the financial assistance of the John S. Guggenheim Memorial Foundation and the National Science Foundation of the United States (grant SES-8908402).

I received very helpful comments and assistance from a number of colleagues both while preparing the lectures and on various drafts of this book; I am especially grateful to Anat Admati, Jeremy Bulow, Drew Fudenberg, Faruk Gul, Frank Hahn and his Quakers, Elchanan Helpman, Paul Klemperer, Marco Li Calzi, Leonardo Liederman, Gerald Meier, Margaret Meyer, Ariel Rubinstein, Tom Sargent, and Eric Van Damme. The opinions and ideas expressed here have resulted from conversations over many years with many colleagues. I cannot possibly give adequate attribution to all, but I would be terribly remiss not to cite the very important contributions of Drew Fudenberg, Faruk Gul, and Robert Wilson to my thinking. (Also, I report in Chapter 6 some of the ideas developed in research I am currently conducting with Fudenberg. I will try to be careful and give appropriate acknowledgements there, but a blanket acknowledgement is called for here.)

Peter Momtchiloff, Andrew Schuller, Anna Zaranko, and many others at the Oxford University Press were always helpful whenever I needed assistance in preparing this book.

I am pleased to dedicate this book to Robert Wilson, my teacher and colleague. Many of the ideas here have been lifted from him, I hope honourably and in a manner that will do him the credit he deserves.

Contents

1
Introduction

Over the past decade or two, academic economics has under-gone a mild revolution in methodology, viz. the language, concepts, and techniques of non-cooperative game theory have become central to the discipline. Thirty-five years ago or so, game theory seemed to hold enormous promise in economics. But that promise seemingly went unfulfilled; twenty years ago, one might perhaps have found 'game theory' in the index of textbooks in industrial organization, but the pages referenced would usually be fairly dismissive of what might be learned from game theory. And one heard nothing of the subject from macro-economists, labour economists . . . , the list goes on seemingly without end. Nowadays one cannot find a field of economics (or of disciplines related to economics, such as finance, accounting, marketing, political science) in which understanding the concept of a Nash equilibrium is not nearly essential to the consumption of the recent literature. I am myself something of a game theorist, and it is easy to overrate the importance of one's own subfield to the overall discipline, but still I feel comfortable asserting that the basic notions of non-cooperative game theory have become a staple in the diet of students of economics.

The obvious question to ask is, Why has this happened? What has game theory brought to the economists' table, that has made it such a popular tool of analysis? Nearly as obvious is, What are the deficiencies in this methodology? Methodological fads in economics are not completely unknown, and anything approaching rational expectations would lead a disinterested observer to assume that the methods of game theory have been pushed by enthusiasts (and especially those who are newly converted) into 'answering' questions that are not well addressed by this particular tool. And third, What next? Will non-cooperative game theory go the way of hula hoops, Davy Crockett coonskin caps, or ———— (fill in the blank with your own favourite example of a fad in economic theory), or will it remain an important tool in the economists' tool-box? Or, set in a way I think is more interesting, In what ways will this tool of analysis change and expand so that it continues to be relevant to economic questions?

In this book, I will give my own answers to these questions. Along the way, I hope to provide enough information about the tools and the questions that these tools can and cannot answer so that you will be able to see what the noise has been about if you have not heretofore followed this mild revolution. (If you are already conversant with the concepts of game theory, you can probably skip Chapter 3 and will probably find my pace throughout to be less than brisk.) If you are able to tolerate abstraction, you may find my manner of providing this information frustrating; I will work with examples and pictures, eschewing formal definitions throughout. If you can stand a formal treatment of the basic ideas and models, consult a proper textbook on the subject.[1]

Now it goes without saying that because I have been something of a protagonist in the rising popularity of game theory as a tool of economic analysis, I have a stake in its continued use by economists. I will claim that game theory's popularity reflects the sorts of things that I have worked on in the past; its weaknesses reflect the sorts of things on which I work today; and its continuing development will reflect the sorts of things on which I hope to be working in the future. Put another way, this is a highly personal statement of the successes, problems, and evolution of game theory in economics.

At the same time, I belong to an intellectual tribe and most of what I will have to say reflects not brilliant personal insights

---

[1] I offer the prejudiced recommendation of Kreps (1990).

but rather the collective view of that tribe, filtered through the lens of my prejudices. Had I written Chapter 6—on where the discipline is or should be heading—five years ago, I might be able to claim some powers of prescience. But my 'predictions' about the future of the subject are very well informed by the current ferment and work in the subject. Any impressive insights that follow should be attributed to the tribe. And if you find yourself thinking what follows is nonsense, just put it down to my poor powers of exposition.

2
The standard

Before setting off, I want to set the standards by which game theory will be judged. For the purposes of this book, *the point of game theory is to help economists understand and predict what will happen in economic contexts.*[1]

I am willing (and able, you may conclude) to twist and turn in all manner of ways the meaning of 'to help us understand and predict'. While it is better if game-theoretic models are not refuted empirically or experimentally, I do not preclude the possibility that one can learn from an empirically falsified model things that help in achieving better understanding and predictions. But I am not interested in this book in game theory as the study *per se* of how ideally rational individuals would interact, if ever they existed. I also will take no notice of how game-theoretic ideas inform fields such as biology. I do not mean to preclude the possibility that economists can learn things of interest and significance to economics from applications of game theory to fields such as biology. But in my opinion we aren't there yet. And I certainly recognize that studying the interactions of ideally rational individuals, if used with circumspection, can aid our understanding of the behaviour of real individuals. But studying the former should aim at understanding the latter. If the theory doesn't help understanding and predictions about real economic institutions and phenomena, then in this book it will be judged a failure.

Having set out such a 'noble' standard,[2] let me begin the

---

[1] I would be happy to extend the domain to include social and political contexts. I believe that most of what I say applies to social sciences other than economics, but this belief is not based on very much first-hand knowledge.

[2] Noble to the social scientist; but perhaps somewhat base to most others.

twisting by noting how it is that understanding and predictions may be aided by the theory.

(1) I take it as an axiom that improvement in understanding by itself implies improvement in predictions. This assertion is hypothetically subject to empirical falsification; ask economists to make predictions about some matter; then expose them to models that they claim improves their understanding of the situation and see if they can make more accurate predictions. But if this empirical test has been carried out, I do not know of it; and so I assume the hypothesis.

(2) Game theory *by itself* is not meant to improve anyone's understanding of economic phenomena. Game theory (in this book) is a tool of economic analysis, and the proper test is whether economic analyses that use the concepts and language of game theory have improved our understanding. Of course, there is an identification problem here. Without the concepts and language of game theory, essentially the same economic analyses may well have been carried out; game theory may be nothing more than window-dressing. Hence improvements in understanding that I may attribute in part to game theory may in fact have little or nothing to do with the theory. Having no ready answer to this criticism, I simply note it and move on. Short of looking for two independent populations of economists, one having game theory at its disposal and the other not, I see no way to solve this problem of identification except by appeal to your good judgement.

(3) Game theory comprises formal mathematical models of 'games' that are examined deductively. Just as in more traditional economic theory, the advantages that are meant to ensue from formal, mathematical models examined deductively are (at least) three: (*a*) It gives us a clear and precise language for communicating insights and notions. In particular, it provides us with general categories of assumptions so that insights and intuitions can be transferred from one context to another and

can be cross-checked between different contexts. (*b*) It allows us to subject particular insights and intuitions to the test of logical consistency. (*c*) It helps us to trace back from 'observations' to underlying assumptions; to see what assumptions are really at the heart of particular conclusions.

So to judge whether game theory helps improve understanding, I will be asking how the theory, employed in the analysis of specific economic questions, provides these advantages in ways that improve our understanding of the specific questions.

In case you missed it, note carefully the elision in the previous sentence. 'In order to judge *whether* the theory helps improve understanding . . .'—which suggests that direct evidence will be brought to bear—'. . . I will be asking *how* the theory . . . provides these advantages . . .'—which suggests a schematic analysis of the question. In order to say definitively whether and how game theory has been successful (or not) as a tool of economic analysis, it is necessary to study in detail game-theoretic analyses of specific problems. The adage that the proof of the pudding comes in the eating applies; to judge whether and why game-theoretic methods have been successful, you must consume their application by reading journal articles and books that apply these methods, such as Tirole's (1988) treatise on Industrial Organization. This book does not provide the substantial fare required for making the judgement. Instead, it resembles a restaurant review, commenting on why and when I (and, I hope, others) have found the pudding tasty; and why and when not. It may help you to understand better what you will and will not be getting with your meal, if you choose to partake. I hope that it whets your appetite. But it is only commentary and not the real thing.

3
Basic notions of non-cooperative game theory

I assume that you (the reader) are familiar with standard economic theory at the level of a good first-year university course. But it is probably unwarranted to assume that you are familiar with the terms of trade of game theory, especially those parts of game theory that have proved so popular recently in economics. So next I will give a few indispensable pieces of terminology and a few examples.

Game theory is divided into two branches, *co-operative* and *non-cooperative* game theory. The distinction can be fuzzy at times but, essentially, in non-cooperative game theory the unit of analysis is the individual participant in the game who is concerned with doing as well for himself as possible subject to clearly defined rules and possibilities. If individuals happen to undertake behaviour that in common parlance would be labelled 'co-operation' (and we will see how this can happen in Chapter 4), then this is done because such co-operative behaviour is in the best interests of each individual singly; each fears retaliation from others if co-operation breaks down. In comparison, in co-operative game theory the unit of analysis is most often the group or, in the standard jargon, the coalition; when a game is specified, part of the specification is what each group or coalition of players can achieve, without (too much) reference to how the coalition would effect a particular outcome or result.

Almost without exception, in this book we will be concerned with non-cooperative game theory only. In the very brief instance when co-operative game theory enters the story, notice will be given.

Strategic form games

There are two basic *forms* or types of formal models that are employed in non-cooperative game theory. The first and more simple is called a *strategic form* or *normal form game*.[1] This sort of model is comprised of three things:

(1) a list of participants, or *players*

(2) for each player, a list of *strategies*

(3) for each array of strategies, one for each player, a list of *payoffs* that the players receive

Since I will try to avoid mathematical definitions, let me give a couple of examples. In the United States (and perhaps elsewhere) children play a simple game called 'rock, paper, scissors'. Two children play at once, so the list of participants is: child A, child B.[2] The two simultaneously choose one of three options; these are the strategies available to each, called: rock, paper, scissors. And depending on what each child selects, the game is either won by one child or the other or it is a draw: If the two choose the same option, the game is a draw. If one chooses rock and the second paper, the second wins. (Paper covers rock.) If one chooses rock and the second scissors, the first wins. (Rock breaks scissors.) If one chooses paper and the second scissors, the second wins. (Scissors cuts paper.) So if we say that the payoff to a child for winning is

---

[1] I will use the term strategic form throughout.

[2] In all the games of this book, the gender of players will be determined by the following general rule: Players A, C, E, and so on will be female, as will players 1, 3, 5,.... Players B, D, F, or 2, 4, 6, will be male. Game theory is one subject where using the two genders is a great help, since writing she and he makes for easier reading than constant reference to players A and B.

Child B

	Rock	Paper	Scissors
Rock	0,0	−1,1	1,−1
Paper	1,−1	0,0	−1,1
Scissors	−1,1	1,−1	0,0

(Child A labels the rows)

FIG. 3.1. The rock, paper, scissors game

1, the payoff for losing is −1, and the payoff for a draw is 0, we can represent this game as in Figure 3.1:

(1) Because there are two players, and each has available three strategies, the set of *strategy profiles* or *strategy arrays* forms a 3 × 3 table.

(2) We list child A's strategies as rows in the table, and child B's strategies as columns.

(3) For each of the nine (3 × 3) cells in the table, we give the pair of payoffs to the two children, first the payoff to child A and then the payoff to child B. So, for example, if child A chooses rock (row 1) and child B chooses scissors (column 3), then child A receives a payoff of +1 and child B a payoff of −1. (Note that row 3, column 1 reverses the payoffs.)

This game is special in many ways, two of which I wish to highlight. First, because it is a two-player game it can be depicted in a two-dimensional table. Second, the sum of the payoffs in each cell is zero. Because of this, this is called a *zero-sum* game. (From the point of view of the theory it isn't important that the sum is zero, but only that the sum is constant, and so instead of zero-sum game the terminology *constant-sum game* is often used.)

Let us look at a second example, in which neither of these special features holds. This game has three players, A, B, and

If player C chooses 1

Player A \ Player B	1	2	3
1	3,3,3	3,2,3	3,1,3
2	2,3,3	2,2,3	2,1,3
3	1,3,3	1,2,3	1,1,3

If player C chooses 2

Player A \ Player B	1	2	3
1	3,3,2	3,2,2	3,1,2
2	2,3,2	6,6,6	6,5,6
3	1,3,2	5,6,6	5,5,6

If player C chooses 3

Player A \ Player B	1	2	3
1	3,3,1	3,2,1	3,1,1
2	2,3,1	6,6,5	6,5,5
3	1,3,1	5,6,5	9,9,9

Fig. 3.2.   A three-player strategic form game

C. Each player has three strategies, namely 1, 2, and 3. Payoffs are simple: Each player receives four times the smallest number that any of them selects as strategy, less the number of the strategy that the particular player selects. For example, if player A chooses strategy 3, player B chooses strategy 2, and player C chooses strategy 3, then the smallest number selected by any of the three is 2. Player A receives $4 \times 2 - 3 = 5$, player B receives $4 \times 2 - 2 = 6$, and player C receives $4 \times 2 - 3 = 5$.

If we wanted to depict this game in the fashion of Figure 3.1, something like Figure 3.2 could be used. Here player A chooses a row, player B chooses a column, and player C chooses one of the three boxes. Think of the three boxes as being stacked one atop another like a multi-storey parking-lot, and then C chooses the level. Payoffs in each of the $3 \times 3 \times 3 = 27$ cells are given first for player A, then B, and then C. Of course, this is far from a constant-sum game; if all three pick strategy 3, then each gets $4 \times 3 - 3 = 9$, for a sum of 27, while if each picks strategy 1, each gets 3, for a sum of 9.

Extensive form games

The second type of model of a game that is used in non-cooperative game theory is an *extensive form game*. In an extensive form game, attention is given to the timing of actions that players may take and the information they will have when they must take those actions. Three examples are given in Figures 3.3, 3.4, and 3.5.

Look at 3.3 first. The picture (and extensive form games in general) is composed of some dots, which are often called *nodes, vectors of numbers, arrows* which point from some of the dots to others and to the vectors, and *labels* for the nodes and for the arrows. Each node is a 'position' in the game; a point at which some player must choose some action. The first position in the game is depicted by an open dot; all the rest are filled in. So, in Figure 3.3, the game begins at the open dot at the top-left of the figure. Notice that each node is labelled with an upper-case Roman letter. These letters index the players in the game—the game in Figure 3.3 has three players, namely A, B, and C—and the letter next to any node gives the identity of the player who must choose an action if that position in the game is reached. So, in particular, at the start of the game in Figure 3.3, it is the turn of player A to choose.

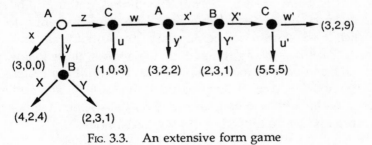

FIG. 3.3. An extensive form game

Coming out of this starting-position or *initial node* are three arrows, labelled x, y, and z. Each arrow represents a choice that is feasible for the player who is choosing. That is, at the start of the game, player A can choose any of x, y, or z. Each of the three arrows points to another position in the game or to a vector of numbers, e.g. if player A chooses y, the next position in the game is the filled-in dot in the middle-left of the figure, which is labelled B (it is B's turn to move) and out of which are two arrows, X and Y. Thus if A starts the game with a choice of y, then B moves next, and B can choose either X or Y. The two arrows X and Y point to vectors of numbers that in this game are composed of three numbers. When an arrow or action points to a vector of numbers, it means that this choice of action ends the game, and then the vector of numbers tells what are the payoffs of the players (it being generally understood that we list payoffs in the order: player A first, B second, and so on). So, to take three examples: if the game starts with A choosing action x, it ends there and A receives 3, B receives 0, and C receives 0; if the game begins with A choosing y, then B moves next, and if B chooses X, then the game ends with payoffs 4, 2, and 4 for A, B, and C, respectively; if the game begins with A choosing z, and then C chooses w, A chooses x', and B chooses Y', the game then ends with payoffs of 2, 3, and 1, respectively.

In these figures two rules are never violated. First, each node has at least one arrow pointing out from it (some action is available to the player) and at most one arrow pointing into it. To be more precise, the initial node has nothing pointing at it; every other node has exactly one arrow pointing to it. Hence if we start at any node except the initial node and we try to backtrack along the path of the arrows, there is always exactly one way to move back unless and until we reach the initial node. Second, if we backtrack in this fashion from any node we never cycle back to the node with which we began; eventually we do indeed get to the initial node.

The effect of these two rules is that our pictures always look like 'trees'; the tree starts at the initial node, and it branches out and out until we reach the end of branches which are the vectors of payoffs, without two branches ever growing into one another and without some branch growing back into itself. In fact, if we expressed these two rules formally mathematically, the name given to the structure that results is an *arborescence* after the analogy to a tree.

Think for a moment about chess and how we would represent it in this fashion. There are two players, white and black. At the opening node it is white's turn to move. White has twenty possible opening moves, viz. advance any one of the eight pawns either one square or two, or move either knight, each to either of two positions. After any of these it is black's turn to move, and so on. There are three things to note about this sort of representation of chess:

(1) When you read in the paper about the 'position' of a chess game, it means the current configuration of the board. There are usually many distinct sequences of moves that lead to a given 'position' so defined. But in our representations of games, a position or node is always defined as a distinct sequence of moves starting from the beginning of the game. Hence a single 'position' in the literature of chess will typically correspond to many different nodes in an extensive form representation of chess; a node is the same as the entire history of moves so far.

(2) Chess, like rock, paper, scissors, is a game that you win, lose, or draw. It is typical to record payoffs for chess as +1 for a win, −1 for a loss, and 0 for a draw. And then the game is a constant-sum game.

(3) Unless you are an *aficionado*, you may think that chess introduces a complication not present in Figure 3.3, namely there may be infinite sequences of moves along which the game never ends. But the rules of chess preclude this. If the

'position' (in the informal sense) is repeated three times or fifty moves pass without a piece being taken or a pawn moved, the game is drawn.[3] Hence an infinite sequence of moves is impossible. We will deal later with games that have infinite sequences of moves, but for now we ignore that possibility.

How could we depict the rock, paper, scissors game with this sort of picture? The difficulty is that the two players move simultaneously or, at least, each must choose between the three options without knowing what the other did. In the pictures we will draw, we depict this game as in Figure 3.4(*a*). As in Figure 3.3, the game begins at the open node or dot. We have labelled this with 'child A', meaning it is child A's turn to choose there. Coming out of this node are three arrows or actions among which child A must choose, viz. rock, paper, scissors. Following each of these we have a (solid) node at which child B must choose between rock, paper, scissors, and then after each of the nine branches so created we give payoffs, first for child A and then for B.

The innovation in this figure concerns the dashed line in Figure 3.4(*a*), which joins the three solid nodes and which is labelled 'child B'. This dashed line is called an *information set*. The idea is that child B, when called upon to choose at any of the three positions or nodes that are joined by the dashed line (that are in the information set) doesn't know which of the three he is at. That is, child B doesn't know what child A selected. Compare with Figure 3.4(*b*) and 3.4(*c*). In 3.4(*b*), we have no dashed lines at all; this represents a game in which child B selects an action after learning what child A has chosen—not a very interesting or fair game. In 3.4(*c*), two of the three nodes are joined in a dashed line; this depicts a situation in which child B learns that child A selected rock if that was A's selection, but if A selects either paper or scissors, then B is only told that the choice was other than rock.

---

[3] I am told that the fifty-move limit may be extended if a player can demonstrate that a forced mate is in progress.

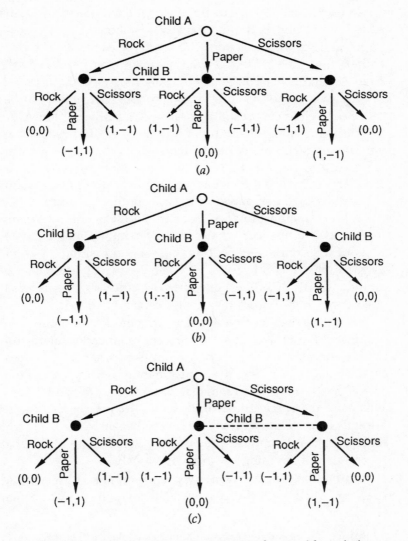

FIG. 3.4.   Rock, paper, scissors in extensive form, with variations

Go back to Figure 3.4(*a*). Suppose that we staged the game as follows: Child B is told to reveal his choice between rock, paper, and scissors to a referee. This referee learns the choice of child B, then goes to child A who sits in another room and asks child A what is her choice *without indicating to her what child B chose*. Child B chooses 'first' in a chronological sense, but because each child chooses in ignorance of the other's choice, the physical chronology is unimportant. Figure 3.4(*a*) remains an entirely adequate model of the situation. What these pictures are meant to represent are (*a*) who are the players, (*b*) what are their options, and (*c*) what they know or don't know about the actions of other players (or, in a moment, nature) when they choose their actions. Actual chronology is important only insofar as it influences what one player knows about the actions of a second. Figure 3.4(*a*) captures the situation as far as game theory is concerned whether the two children choose simultaneously or (either) one after the other, as long as when choosing neither knows the other's choice.[4]

Information sets can be used in game models of the sort we are describing to depict many different manners of interaction among the players. But there are limits to what is permitted. First, if two nodes are in an information set, then the same player has the choice to make at those nodes, and the player will have the same range of choices. Second, in the games we will discuss, players will have good memories. Each will remember that he moved previously (whenever he did), what he chose, and anything he knew previously.[5]

Information sets and what they represent take some getting used to. As noted already, we will not launch into formal

---

[4] Perhaps you are unconvinced that physical chronology will be irrelevant in all cases. I am unconvinced of this myself and will return to this point in Chapter 5. Standard game theory ignores chronological variations that do not affect informational conditions, but as we shall see standard game theory does not take into account many factors of consequence to how individuals play games.

[5] For those readers who have seen a formal treatment of the subject before, let us paraphrase: We will only deal with games of perfect recall in this book.

definitions here, but instead work by means of examples. So let me close with one further example—the *truth game*—that introduces yet another element of extensive form games.

Imagine I have a coin which is bent so that when flipped, it comes up heads 80 per cent of the time. I, as something of a referee, set out to play the following game with two players, whom I will call S (her) and R (him). First I flip the coin. I note the result and reveal it to S. Then S must make an announcement to R; S may say either 'Heads' or 'Tails'. R hears this announcement and then must respond with either 'heads' or 'tails'. Note well, R hears what S said but doesn't see what S saw; R doesn't know the results of the actual coin-flip when he makes his response. Payoffs to the two players are as follows: R gets 10c if his response matches the results of the actual coin flip and he gets nothing otherwise. S gets 20c if R says 'heads' and she gets a further 10c if she tells the truth about the coin-flip in her announcement.

An extensive form depiction of this situation is given in Figure 3.5. It works as follows. The game begins with the open dot in the centre of the figure, which is labelled N and which has two arrows emanating from it, one marked H and one T. This 'choice' represents the results of the coin-flip, and the node is labelled N because this particular choice is a result of the forces of Nature. This is how we deal with games in which some elements are random. We imagine a 'player' called Nature who makes these choices. Nature will not have payoffs, but she does have positions or nodes under her control, and she chooses (at the appropriate instances) among the 'actions' that emanate from her nodes. It is conventional to assume, and we will do so throughout, that all the players in the game share a common probability assessment concerning the odds with which Nature makes her choices; these are depicted by probabilities in curly brackets on the appropriate arrows, such as the 0.8 and 0.2 in Figure 3.5.[6]

---

[6] While this assumption is conventional, it is not necessary.

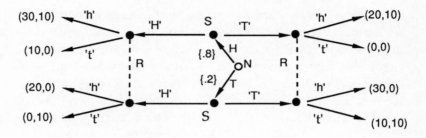

FIG. 3.5.   The truth game in extensive form

Now suppose Nature chooses H. We move to the solid node in the top-centre of the figure, where it is S's turn to choose between 'H' and 'T'. And if Nature chooses T, we move to the solid node in the bottom-centre of the figure, where it is again S's turn to choose between the same two options. Note the absence of an information set; S is told the result of the coin-flip.

Suppose Nature chooses H and S chooses 'H'. We are at the node in the top-left quadrant of the figure. It is R's turn to choose, between 'h' and 't'. The crucial thing to note here is that this node is joined by a dashed line to the node in the bottom-left quadrant of the figure; and similarly, the two nodes on the right-hand side are joined by a different dashed line. The dashed line on the left, labelled with R, indicates that R chooses at this information set, an information set that consists of the position of the game after the sequence of actions H-'H' and the position after the sequence T-'H'. R knows the announcement made by S; thus there are two information sets for R, one on the left and one on the right. But R doesn't know what Nature did; R's state of information after the sequence H-'H' is precisely the same as his state of information after T-'H', namely that S sent the message 'H'.

Finally, payoffs are made to the two players, first S and then R, in accordance with the rules of the game.

We will see examples of extensive form games throughout

this book, so if the rules are still fuzzy they should clarify as we go on.

Extensive and strategic form games

We now have two different sorts of models of a 'game', the strategic form model and the extensive form model. What are the connections between the two? For our purposes, the following statement sums up the matter.

*To every extensive form game there is a corresponding strategic form game, where we think of the players simultaneously choosing **strategies** that they will implement. But a given strategic form game can, in general, correspond to several different extensive form games.*

To understand this statement you must understand the notion of a player's *strategy* within the context of an extensive form game. Consider the truth game as depicted in Figure 3.5. Imagine that player S (say) wishes to leave explicit instructions to some agent concerning how she wishes the agent to play in her behalf. Player S must leave instructions what do to if Nature chooses H *and* what to do if Nature chooses T. Think in terms of a menu in which you have one choice of entrée out of two and one choice of dessert out of two; from this you can compose four possible meals. And so it is in this case; S can compose four possible strategies for her agent, namely

s1: If Nature chooses H, say 'H'; if nature chooses T, say 'H'

s2: If Nature chooses H, say 'H'; if nature chooses T, say 'T'

s3: If Nature chooses H, say 'T'; if nature chooses T, say 'H'

s4: If Nature chooses H, say 'T'; if nature chooses T, say 'T'

Each of these is a strategy for player S, a complete specification of how she will act in every choice situation in which she might find herself.[7]

---

[7] The analogy to a menu is not perfect. Presumably you will have both an

And at the same time R has two choice situations, his two information sets, and in each he has two possible actions, for 2×2 possible strategies. These are

r1: If S says 'H', respond 'h'; if S says 'T', respond 'h'

r2: If S says 'H', respond 'h'; if S says 'T', respond 't'

r3: If S says 'H', respond 't'; if S says 'T', respond 'h'

r4: If S says 'H', respond 't'; if S says 'T', respond 't'

Now consider the strategic form game depicted in Figure 3.6 and think about the following story. S and R are unable to play this game; each will be out of town when the coin will be flipped. Each is allowed to appoint an agent to play in their (respective) steads; each must leave explicit instructions for their agent. S chooses her instructions from the list s1 through s4; R simultaneously and independently chooses his instructions from the list r1 through r4. So we have the two picking strategies simultaneously and independently, each from a given list of strategies, which is just the story of a strategic form game.

It remains to explain the payoffs in Figure 3.6. Suppose S chooses strategy s2 and R chooses strategy r3. What actually happens depends on the coin-flip. If the coin comes up heads, then strategy s2 calls for S (or her agent) to announce 'Heads', and then strategy r3 calls for R (or his agent) to respond 'tails'. Look back at Figure 3.5, upper-left quadrant, and you will see that this gives 10c to S and zero to R. Whereas if the coin flip results in tails, s2 calls for S to announce 'Tails' and r3 (then) calls for R to respond 'heads', which nets 30c for S and zero for R. Since the first possibility has probability 0.8 and the second 0.2, the expected payoffs from this pair of strategies are $(0.8)(10) + (0.2)(30) = 14c$ for S and $(0.8)(0) + (0.2)(0) = 0c$ for R; look at the corresponding cell in the strategic form game in Figure 3.6. The other pairs of payoffs are computed similarly.

---

entrée *and* dessert. But in the case of the truth game, S will be told either H *or* T. Think of a restaurant where you order both an entrée and a dessert when you enter, but then the kitchen decides whether to give you one or the other.

Player R's strategy

		r1	r2	r3	r4
	s1	28,8	28,8	8,2	8,2
	s2	30,8	26,10	14,0	10,2
	s3	20,8	4,0	16,10	0,2
	s4	22,8	2,2	22,8	2,2

Player S's strategy

Fig. 3.6.   The truth game in strategic form

We just glided past one technical point and without making a big deal out of it, you are owed an explanation. When we compute payoffs and the game has moves by nature, we will use expected payoffs in the manner just employed. We can justify this in two ways. First, it may be that the players regard lotteries in just this fashion; one lottery is better than another if it has a higher expected payoff. (To use jargon, the players are risk-neutral in the prizes.) Or, more technically, we can refer to one of the standard theories of choice under uncertainty where one derives conditions under which individuals choose according to the 'expected utility' of the prizes they might receive in various lotteries; we then assume that those conditions hold and that payoffs are computed in terms of utility. (If you know about models of choice under uncertainty and expected utility, this second justification will make sense; if not, you must make do with the first justification, which is probably sensible enough if the range of possible prizes is small.)

Is the extensive form game in Figure 3.5 the 'same' as the strategic form game in Figure 3.6? Our story of the two players instructing agents would seem to make it so, at least intuitively. But if we do without the artifice of these agents, are the two games really the same? This is a difficult question; one on which hangs a fair bit of debate among game theorists. We will proceed for now on the following basis: The two are different representations of the same game, and we choose which we want to use according to convenience in representation or analysis. But when we revisit this matter briefly in later chapters, you will get a sense of the debate.

Having begged this question, the example shows how one can get from an extensive form game to a strategic form game. The set of players doesn't change. The set of strategies for each player corresponds, menu fashion, to an array of choices made by the player, one choice for each information set controlled by the player in question. And given an array of strategies, one for each player, we are able to work through to the payoffs each player gets, averaging across any action by Nature.

Going in the other direction, from a strategic form game to an extensive form game, is not so clear-cut. For any strategic form game we can immediately construct a corresponding extensive form game in which the players simultaneously and independently choose strategies. That is, the two-player strategic form game depicted in Figure 3.7(a) can be turned into the extensive form game depicted in Figure 3.7(b); player A chooses one of her two strategies, and player B simultaneously and independently chooses one of his two. Note that our repeated use of the phrase 'simultaneously and independently' connotes the information set of player B; player B chooses without knowing what player A has done. Alternatively, we can use the extensive form game given in Figure 3.7(c); here all we've done is to interchange the 'chronological order' of A and B, having B choose 'first'. But, by virtue of the information set of A, A doesn't have any information about B's choice when she chooses.

A more interesting extensive form game corresponding to the strategic form game in Figure 3.7(a) is given in Figure 3.7(d). Here player A moves first, and player B certainly knows what

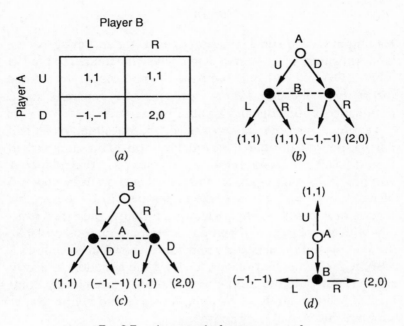

FIG. 3.7.   A strategic form game and
three corresponding extensive form games

player A did when he moves, since if A chooses U, player B doesn't move at all! But if you consider the matter carefully, you will see that the extensive form game in Figure 3.7(*d*) does indeed correspond to the strategic form game in Figure 3.7(*a*), by the rules for correspondence that we gave before.

In general, then, strategic form games can correspond to a number of different extensive form games. *If* we believe that all the relevant information about the game is encoded in the strategic form game model, then we are implicitly saying that there is no essential difference between the extensive form games in Figures 3.7(*b*), (*c*), and (*d*). (What do you think about that hypothesis in this case?)

Dominance

Having modelled some situation as a strategic form or extensive form game, the next step is to analyse the model, to predict what will (and will not) happen. In non-cooperative game theory two so-called *solution techniques* are generally used, dominance arguments and equilibrium analysis.

Dominance is one way to answer the question, What will not happen? Consider the strategic form game depicted in Figure 3.8(*a*). It seems reasonable to suppose that player A will not choose strategy x because whichever strategy player B picks, strategy y gives player A a higher payoff. Of course, this argument is based on the presumption that A picks her strategy with the objective of getting as high a payoff as possible, but that presumption is fairly non-controversial in economics.[8] Also, it presumes that player A views the situation in a way similar enough to our model so that A recognizes that y dominates x. But given these presumptions it is entirely natural to suppose that A will not choose x.

In Figure 3.8(*a*), we can predict that player A will not choose x, but the logic that led us to this prediction takes us no further. Compare with the game in Figure 3.8(*b*). Player A's payoffs have not been changed, so once again we say that x is dominated by y for A, hence it is unlikely that A will choose x. But now we can carry this a step or two further. If B comes

---

[8] At the risk of unnecessarily confusing you, let me add that in a sense this assumption is tautological. Payoffs are meant to represent numerically the players' preferences, which in turn are meant to 'represent' their choice behaviour. So if we see a player choosing in a fashion that doesn't maximize his payoffs as we have modelled them, then we must have incorrectly modelled his payoffs. We will return in chapter 5 to this point and to the question, What are a player's payoffs? Note also that even if it is tautological that players maximize their payoffs, this doesn't quite imply that they avoid dominated strategies. To come to the second conclusion we must further assume that each player believes his choice of strategy does not affect the choices by others. This may seem entirely reasonable, but I have met students (playing the prisoners' dilemma game, to be described later) for whom this assumption failed.

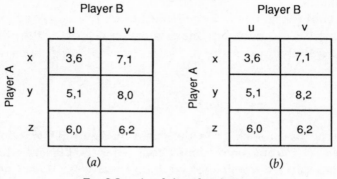

Fig. 3.8. Applying dominance

to the conclusion that A is unlikely to play x, *then* B's strategy u is dominated by v. Note carefully that if A chooses x, then u is better than v for B. But against either y or z, v is better than u. Hence once we (and, more to the point, once B) decides that x will not be played, then we can predict that u will not be played. The implicit presumptions in this argument are more problematic than before. Now we not only presume that A will (probably) avoid a strategy that gives a worse payoff no matter what B does and that A views the game as in the model, but also that (*a*) similar assumptions are true about B, and (*b*) B makes these assumptions about A. It is the second new assumption that deserves scrutiny; we are making an assumption about what one player thinks about others.

In the case of the game in Figure 3.8(*b*) we can carry this a step further. Since v dominates u (after x is removed from consideration), we make the prediction that B will not choose u. (In this game, this means that we predict that B *will* choose v, but the logic of dominance should be thought of as logic that tells you what *won't* happen.) If A makes a similar prediction based on the assumption that B makes the assumptions above, *then* y dominates z, and we can predict that z won't be played. (Hence our prediction is that A will choose y and B, v.)

The first step in chains of logic of this sort involves the application of *simple dominance*. When we go back and forth, first eliminating one or more strategies for one player and then on that basis eliminating one or more strategies for the others, we are applying *recursive* or *iterated* or *successive dominance*.

Nash equilibrium

In some cases, such as in the game in Figure 3.8(*b*), we can offer fairly strong predictions about what the players will do if we apply the dominance criterion iteratively. But in many other games that will be of interest, dominance gets us very little. In this case it is typical in the literature to resort to *Nash equilibrium analysis*.

A Nash equilibrium is an array of strategies, one for each player, such that no player has an incentive (in terms of improving his own payoff) to deviate from his part of the strategy array. For example, in the game in Figure 3.2 it is a Nash equilibrium for player A to choose the second row, player B to choose the second column, and player C to choose the second box. To see why, note that if player B is choosing the second column and player C is choosing the second box, then player A gets 3 by choosing the first row, 6 by choosing the second row, and 5 by choosing the third. Hence the second row is player A's best response to what the others do. Similarly the second column is B's best response to the choice of the second row by A and the second box by C, and the second box is C's best response to the second row by A and the second column by B. That is the criterion of a Nash equilibrium; each player is maximizing on his own, given the supposed actions of the others.

Note the strong non-cooperative flavour of this definition. Consider, for example, the game depicted in Figure 3.9. This famous game is known as the *prisoners' dilemma* for reasons we will explain in Chapter 4. For now, note that in this game row 2 combined with column 2 is a Nash equilibrium. (Indeed, row 2 dominates row 1 and column 2 dominates column 1.) But if

Prisoner B

		Column 1	Column 2
Prisoner A	Row 1	5,5	−1,6
	Row 2	6,−1	0,0

Fɪɢ. 3.9.   The prisoners' dilemma

players A and B decided together to change their actions, each would be better off. Recall at this point that the subject is non-cooperative game theory. If we thought that players A and B could somehow come to a binding agreement for A to choose row 1 and for B to choose column 1, then we would include that possibility in the game; either we would have a strategic form game with strategies that represent actions such as 'Sign a contract that obligates you to choose row/column 1 if the other party signs a similar contract' or we would include in an extensive form representation the possibility of similar actions. When we write the game as in Figure 3.9 and we proceed to analyse it with the concepts of non-cooperative game theory, the notion is that such possibilities are not present, players must choose their actions simultaneously and independently, and (thus) each will probably choose the 'non-cooperative action', since there is no way to bind the other party or for the other party to inflict on the first some form of punishment.

A given game may have many Nash equilibria. The game in Figure 3.2 is a good example. In this game, row 2, column 2, box 2 is one Nash equilibrium. But row 1, column 1, box 1 is another; and row 3, column 3, box 3 is a third.

This immediately suggests a question. In the analysis of economic institutions, the analyst will pose a game-theoretic model and then, very often, identify a Nash equilibrium as 'the solution', i.e. the analyst's prediction of what will ensue in the situation being modelled. What is behind this peculiar practice, especially in cases when the game in question has multiple Nash equilibria?

No fully satisfactory answer to this question can be offered or, rather, no answer can be offered which will satisfy all practitioners of game theory in economics. Indeed, many of the controversies concerning the application of game theory to economics revolve around this question, to which we will return in Chapters 5 and 6. I will develop my own answer to this question throughout this book, and it will be evident that the answer that I begin to give here colours many of the conclusions to which I will later come; economists and game theorists who disagree with my later conclusions will trace their disagreement from this point.

Remember that the purpose of the theory is to help us to understand and predict economic phenomena. When the dominance criterion is applied, it is under the tacit assumption that individuals will not choose strategies that are dominated; when iterated dominance is applied, it is under the tacit assumption that players in a game will see and act upon the hypothesis that others will not play dominated strategies, and so on. Insofar as these tacit premises are incorrect (and in chapter 5 we will see a situation in which this *seems* to be the case), the application of dominance or iterated dominance in predicting what will happen in a particular situation will be misleading. But insofar as these tacit premises are correct, dominance gives us a very clean and direct engine for making predictions.

With Nash equilibrium, the 'logic' is a good deal more tortured. I assert that in *some* situations all the participants are fairly clear in their own minds what they and others should do. In such cases it is necessary that this 'evident' course of action constitutes a Nash equilibrium; otherwise one or more of the individuals involved would see some other course of action as being in his own best interests. For example, consider the

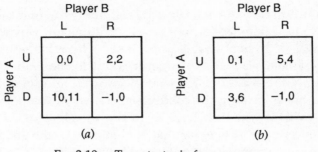

Fɪɢ. 3.10. Two strategic form games

strategic form game in Figure 3.10(*a*). Most people considering that game from either the position of player A or player B will understand that the called-for actions are D for player A and L for player B. The reasons why are clear *in this instance*; it is in the players' joint interests. What we are saying is that in such a case (and, in particular, in this case), if a mode of behaviour is self-evident, and each player believes it is self-evident to all the players, it is self-evident that it is self-evident, and so on, then each of the players must be choosing a best response to what the others are evidently doing. That is, it must be a Nash equilibrium.

This is not to say that every game situation has a self-evident way to play. The game in Figure 3.10(*b*), played by two individuals who have no opportunity to communicate beforehand and have had no prior experiences with each other, has no self-evident way to play. (It does have Nash equilibria, of course; D-L is one and U-R is another.)

Indeed, one can cook up games with a single Nash equilibrium that is not the self-evident way to play, and one can cook up games in which the majority of play involves strategies that are not part of a Nash equilibrium. (These examples will be given in Chapter 6.) Unless a given game has a self-evident way to play, self-evident to the participants, the notion of a Nash equilibrium has no particular claim upon our attention.

Hence when economic analysts invoke the notion of a Nash

equilibrium, they are asserting at least implicitly that the situation in question has (or will have) a self-evident way to play. When, as often happens, they don't say why it is that there is a self-evident way to play, then it is left to the reader either to supply the reason or to be suspicious of the results. Either the author is making the conditional statement that *if* there is a self-evident way to behave (or if one arises), *then* the equilibrium mode of behaviour that is identified in the analysis is a good prediction what that way to behave will be. Or the author is asserting the joint hypothesis that there is a self-evident way to behave and it is the equilibrium that is identified, and the author isn't defending the first part of this joint hypothesis.

Please note carefully the two parenthetical insertions in the previous paragraph. A common justification for Nash equilibrium analysis holds that such analysis is useful for making predictions in cases where players can gather beforehand for pre-play negotiation. They sit down together and try, via some procedure, to reason through how each should act. Whether they come to agreement or not, they eventually play the game; *if* they come to agreement, and if that agreement is credibly self-enforcing in the sense that no one who believes that others will conform has the motivation to deviate, then the agreement will be a Nash equilibrium. Put the other way around, the set of Nash equilibria contains the set of credibly self-enforcing agreements that could be made. When Nash equilibrium analysis is justified by this story, then the analyst is saying something like, 'I believe that pre-play negotiation will result in a self-enforcing agreement, and by identifying Nash equilibria I identify the range of possible agreements.' Of course, when the analyst goes on to pick one or several of many such equilibria on which to concentrate attention, then the analyst is maintaining that he knows something about the nature of the agreement that will be reached.

If we could justify Nash equilibria only by the story of pre-play negotiations, then the power of the concept for studying economic phenomena would be low. There are many situations in which participants do sit down to negotiate before actions are taken, but there are more in which this sort of explicit negotiation does not take place. For example, a very popular context for applying game-theoretic notions has been in industrial organization, studying the behaviour of oligopolists in a given industry. At least in the United States, if oligopolists sit down together to negotiate how each should behave, and if the Justice Department learns of this, someone could be sent to jail. So to extend the reach of the concept we look for other reasons that a self-evident way to behave may exist in particular situations.

The game in Figure 3.10(a) gives one situation that even absent pre-play negotiation seems to most individuals to suggest a self-evident way to play. And, as we said, in this instance it is the commonality of interests that makes it so. Here is another, more complex game in which most people know what to do. Consider the following list of nine cities: Berlin, Bonn, Budapest, London, Moscow, Paris, Prague, Warsaw, Washington. Suppose two players are asked each to make a list of these cities on a piece of paper, without consultation. Player A's list must include Washington; and the B's must include Moscow. If the two lists exactly partition the set of nine cities (each city appears on one and only one list), each player gets $100. Otherwise, each gets $0. How likely do you think it is that the two will win the $100? What do you think will be the two lists?

I have never tried this game on subjects, but I feel safe in guessing that the players would do very well indeed.[9] The list with Moscow will normally include Berlin, Budapest, Prague, and Warsaw, and the list with Washington will have Bonn, London, and Paris. The rule of division is evident; capitals Warsaw Pact countries vs. those of NATO countries (abusing, for the moment, French sensibilities). Berlin might give one

---

[9] As I write this, events in Europe may be rendering this example obsolete.

pause, but the presence of Bonn on the list is something of a give-away, and, well, it just seems fairly clear.

Imagine, on the other hand, that the list was Berlin, London, Moscow, Rome, Tokyo, and Washington, with Berlin specified as being on one list and Washington on the other. Do you think Axis vs. Allied Powers? East vs. West? Pacific-rim nations vs. strictly European (now perhaps offending some British sensibilities)? It isn't so clear this time.

Or go back to the original list and imagine a more complex game. Now each city has been assigned a number of points, obtained by a complex formula involving the artistic and cultural importance of the city. The points range from 100 points to 60; Paris is highest with 100 points, and Bonn lowest with 60. The players are not told these point-values. Again the two players are asked to make independent lists of cities, but now prizes are more complex. For every city appearing on one list and not the other, the person who listed the city gets as many dollars as the city has points. For every city appearing on both lists, both lose twice the number of dollars as the city has points. And if the two players partition precisely the list of nine cities, prizes are doubled. Before, the two players had precisely coincident objectives, to partition the list of cities. Now they want to partition the list, but each would like his portion of the partition to be as large as possible. Even so, I would bet that tried on reasonably intelligent university students, this game would result in the Warsaw Pact–NATO plus France partition in a majority of trials. (I have played a similar game with American students, using American cities divided by the Mississippi River, and my rate of success predicting the partition is around 80 per cent.)

In certain situations, participants do seem to 'know' or at least have a good idea how to act. From where does this knowledge come? If we imagine two (or more) individuals interacting repeatedly, then *modi vivendi* may develop between

(or among) them through a process of trial and error.[10] Among
individuals who have never faced each other, if they have been
in similar situations before they may have learned social conventions about how to act and what actions to expect in return. And even individuals who have never faced each other
and who have never been in a situation very much like the
one they now face may be able to see in the current situation
cues that they think will be suggested to all concerned parties.
Those cues (and their existence) may depend on the identities
of the parties; if you were playing the cities game against an
American grade school student, you might have less faith in
your opponent's ability to split European capitals into Warsaw Pact capitals and their complement. But in many situations those cues are present and suggest how to proceed. It
is here that the notion of Nash equilibrium enters; if the cues
are meant to suggest themselves to all parties concerned, then
they should suggest a mode of behaviour that constitutes a
Nash equilibrium.

It is, of course, a long way from these toy games to situations
of economic relevance. But economic contexts, or rather some
economic contexts, are perhaps more apt to have 'conventional
modes of behaviour' than are cute and made-up games involving the capital cities of Europe. As long as such conventions
suggest to all parties what their role is to be, and as long as
parties obey conventions in their own self-interest as defined
by the payoffs given by the game model, then conventional
behaviour should be a Nash equilibrium. And analysing the
situation using the notion of a Nash equilibrium will give an
outer bound to the set of possibly 'stable' conventions.

You may at this point be scratching your head and won-
dering what sort of mess all this is, if we get to such muddy

---

[10] As we shall see in Chapter 4, two or more individuals who interact repeatedly are typically not bound by the dictates of a static equilibrium. If this
remark makes no sense now, don't worry; it is there to cover my last remark
for the benefit of readers who already know some of what is coming.

philosophical issues in a chapter that is called 'basic notions'. So we leave this issue at this point, to see some of the things that game theory contributes to economic thinking. But you can be sure that we will return to these issues later, since the very relevance of Nash equilibrium analysis is at stake here, and Nash equilibrium analysis makes up a lot of what goes on in game theory applied to economics. Indeed, to anticipate Chapters 5 and 6, the great weaknesses of game theory are that it is fuzzy (to say the least) on just when and, if so, why equilibrium analysis is relevant, and on what to do when equilibrium analysis is irrelevant. Attempting to deal with that fuzziness is, in my opinion, a crucial direction for future research in the subject, if it is to continue to improve our understanding of economics.

4
The successes of game theory

In this chapter we give answers to the question, What have been the successes of game theory in economic analysis? At the outset let me remind you that I will be dealing almost entirely with non-cooperative game theory. So I will ignore the many contributions co-operative game theory has made, especially to the theory of general equilibrium.

Taxonomy based on the strategic form

The first contribution made by game theory goes back to some of the earliest developments of the theory (and is therefore not really part of our story). Game theory provides a taxonomy for economic contexts and situations, based on the strategic form. An example is the prisoners' dilemma game, depicted in Figure 3.9 and reproduced here (slightly modified) as Figure 4.1. The story that gives this game its name runs as follows. The police have apprehended two individuals whom they strongly suspect of a crime (and who in fact committed the crime together). But the police lack the evidence necessary to convict and must release the two prisoners unless one provides evidence against the other. They hold the two in separate cells and make the following offer to each:

> Implicate your colleague. If neither of you implicates the other, each of you will be held for the maximum amount of time permitted without charges being made. If one of you implicates the other and is not implicated, we will release the first and prevail upon the judge to give the recalcitrant second party the maximum sentence permitted

by law. If both of you implicate the other, then both will go to gaol, but the judge will be lenient in view of your co-operation with the authorities.

The story suggests that of the four possible outcomes for a prisoner, it is best to implicate and not be implicated, second best neither to implicate nor to be implicated, third (and a good deal worse) to implicate and be implicated, and worst to be implicated while failing to implicate your colleague. This general ordering of payoffs is captured by the numbers in Figure 4.1. (You should have no trouble seeing that in the absence of other considerations of the sort we will discuss later in this chapter, we are led to predict that each side will implicate the other, since this is a dominant strategy for each.)

This general sort of situation recurs in many contexts in economics. Consider two firms selling a similar product. Each can advertise, offer items on sale, and so on, which may improve its own profit and hurt the profits of its rival, holding fixed the actions that its rival may take. But increased advertising (and so on) by both decreases total net profits. (If you like to think of concrete examples, consider the competition between Airbus and Boeing to sell airplanes. Although advertising is not an important sales variable here, price concessions are, and major airlines often do very well by playing one of the two airframe manufacturers against the other.) Each of the two firms is led to advertise in order to increase its own profits at the expense of its rival's, and if both do so, both do worse than if they could sign a binding agreement (or otherwise collude) to restrict advertising. Although more complex than the prisoners' dilemma game because the decision to be taken is presumably more complex than a simple advertise/don't advertise binary choice, this situation is 'strategically in a class with' the prisoners' dilemma, and we can hope that by studying how players might interact in the stark prisoners' dilemma game we can gain insight into the rivalry between the two firms.

Or consider two countries that are trading-partners. Each

Prisoner B

		Don't	Implicate compatriot
Prisoner A	Don't	5,5	−1,6
	Implicate compatriot	6,−1	0,0

Fɪɢ. 4.1. The prisoners' dilemma

can engage in various sorts of protectionist measures that, in some cases, will benefit the protected country, holding fixed the actions of the other. But if both engage in protection, overall welfare in both countries decreases. Again we have the rough character of the prisoners' dilemma game, and insights gained from, say, the context of rivalrous oligopolists might be transferable to the context of trade policy. To take a third example, imagine two tax jurisdictions which compete for industrial development by offering tax concessions. The list can go on for quite a while; the point is that the stark structure of the prisoners' dilemma game can be 'found' in all sorts of contexts, which can then be so characterized and examined for similarities.

The prisoners' dilemma is, perhaps, the most prevalent example of a basic game that recurs (in vastly more complex form) in economic contexts, but it is not the only one. A second example is the so-called *battle of the sexes*. The story concerns a husband and wife who are trying to decide how to spend an evening. The man would like to attend a boxing match; the woman, the ballet.[1] But each would rather attend an entertainment with the other than not, and this is more important to each than attending his or her preferred form of entertainment. This suggests the payoffs depicted in Figure 4.2.

[1] Classic examples in game theory were developed many years ago, and I hope the reader will not be put off by the sexist elements.

Husband

		Go to boxing	Go to ballet
Wife	Go to boxing	4,5	0,0
	Go to ballet	1,1	5,4

FIG. 4.2.   The battle of the sexes

This game is representative of many situations in which two (or more) parties seek to co-ordinate their actions, although they have conflicting preferences concerning which way to co-ordinate. The 'cities' game of Chapter 3, where payoffs to a given player are greater the more cities that player lists, is of this character. In the context of industrial organization, market segmentation among rivals can have this flavour, as can situations where two manufacturers of complementary goods contemplate standards to adopt; they wish to adopt compatible standards, but for various reasons each may prefer a different sort of standard. In the context of public finance, two adjacent tax authorities may wish to co-ordinate on the tax system they employ to prevent tax-payers from benefiting from any differences through creative accounting procedures, but if each has a distinct tax clientele, they may have conflicting preferences over the systems on which they might co-ordinate. (For this situation to be similar to the battle of the sexes, it is necessary that each finds it more important to match the tax system of the other party than to tune its tax system to its home clientele.) In the context of labour economics, both a labour union and management may be better off settling on the other's terms than suffering through a strike, but still each will prefer that the other party accommodate itself to the preferred terms of the first.

And so on. This is not the place, nor do we have the time,

to give a complete description of the taxonomy of strategic form games. But these two examples should suffice to make the point: Although such taxonomies may miss many crucial details in the particular application, they do capture basic strategic aspects of various situations, allowing economists to make linkages between applications.

Dynamics and extensive form games

You may think that in all of the applications given, the strategic form taxonomy is too weak to be of much use because all these situations have a dynamic aspect that the simple strategic form games miss. If you think this, then your thoughts are similar to the thoughts of many economists who regarded the modelling and analysis of economic situations in terms of strategic form games as wild (and largely useless) over-simplification. The recent impact of game-theoretic methods in economics, I contend, traces in large measure from the ability to think about the dynamic character of competitive interactions, by using extensive form games. Let me put this in italics because it is one of the central themes of this chapter: *The great successes of game theory in economics have arisen in large measure because game theory gives us a language for modelling and techniques for analyzing specific dynamic competitive interactions.*

Let me illustrate with an example taken from the literature on industrial organization, the theory of an entry-deterring monopolist, due to Bain (1956) and Sylos-Labini (1962). We imagine a monopolist (she) in the classic sense; a manufacturer, say, who alone produces a good for sale. This monopolist faces a downward-sloping demand curve. For simplicity, we imagine that if the monopolist sets price p, then the quantity demanded x is given by $x = 13 - p$. This monopolist has a simple cost structure as well; the total cost of producing x units, for $x > 0$, is $x + 6.25$. That is, there is a fixed cost of 6.25 and a constant marginal cost of 1 per unit.

The standard theory of monopoly would proceed as follows. If the monopolist produces x units, her total revenue will be $(13 - x)x$, and so total profits will be

$$(13 - x)x - x - 6.25 = 12x - x^2 - 6.25.$$

To maximize this in x means setting the derivative of total profits to zero, or $12 = 2x$, or $x^* = 6$, which (working back) gives total profits of 29.75. However, the theory of entry deterrence goes, we might instead observe the monopolist producing not $x^* = 6$ but instead $x^0 = 7$ units, which gives total profits of 28.75, with the following explanation:

This monopolist is faced with the threat of entry by some second firm, known as the entrant (he). This entrant is as capable a manufacturer as the monopolist; he possesses exactly the same production function as does the monopolist. And he will enter the industry if he believes that he can make a strictly positive profit by doing so. Suppose that the monopolist is producing $x^* = 6$ units, *and the potential entrant believes that if he enters, the monopolist will go on making 6 units.* Then the entrant, if he plans to produce y units, will obtain a price of $13 - 6 - y$ for each unit he sells, for net profits of

$$(13 - 6 - y)y - y - 6.25.$$

This is maximized at $y^* = 6/2 = 3$ for net profits of 2.75. This is strictly positive, and the entrant will enter. (If we believe with the entrant that the monopolist will continue to produce 6 units and the entrant 3, this means that the monopolist will net 11.75 in profit following entry.)

If, on the other hand, the monopolist is producing $x^0 = 7$ units, and the entrant believes that if he enters, the monopolist will go on making 7 units, then the entrant will obtain profits of

$$(13 - 7 - y)y - y - 6.25$$

by producing y units. This quantity is maximized at $y^0 = 5/2$ for net profits of precisely zero. This is just insufficient to warrant entry (since the entrant is attracted to the industry only if he sees positive profits), and so the entrant chooses not to enter.

Thus the monopolist is justified in producing $x^0 = 7$ units instead of $x^* = 6$. It is true that this costs the monopolist one unit of profit, relative to what she could have if she produced the 'myopic' optimal quantity of 6. But producing 7 units forestalls entry, and preserving the monopoly leads to greater profits in the long run than does extracting the last penny out of the market in the short run and suffering from the competition that would result.

Or so runs the story. For those readers unaccustomed to the ways in which economic theorists work and think, this will seem a fairly stark story. But it is meant to capture the essence of the phenomenon of entry deterrence, wherein a monopolist maintains her stranglehold on a market by taking actions that forestall entry by potential competitors. The monopolist in so doing sacrifices short-run profits, but she thereby secures greater long-run profits. When studying and making predictions about monopoly industries, this simple model teaches us to watch for the source of the monopoly power and to temper the predictions we make about how the monopolist will act if monopoly power is not legally mandated but is maintained by some sort of deterrence strategy.

For readers perhaps a bit more accustomed to thinking about industrial conduct, this story will still seem a bit stark on somewhat different grounds. *In this story or in any situation that this story is meant to represent, why does the entrant believe that the monopolist will continue her pre-entry production levels if and when entry does occur?* These conjectures by the entrant (and the monopolist's comprehension of them) are essential to the story, and at the same time they are unexplained by the story. We could simply assume that entrants hold such conjectures or, rather, qualify the story by saying that when entrants hold such conjectures, then entry-deterring behaviour by a monopolist might be observed. But we might also seek to analyse the situation further to see what might explain or justify such conjectures as these (or to see why such conjectures, and hence the entire theory, might be utter nonsense).

A seemingly likely path for deeper analysis involves studying the details of the competitive interaction between monopolist and entrant. The story told above suggests an element of time or dynamics: The monopolist chooses a production quantity to which the entrant responds. But the detailed structure of the dynamics is not explicit. Perhaps the monopolist somehow can commit herself to a production quantity which she must supply whether there is entry or not. Perhaps such commitments cannot be made, but the monopolist is able to take actions pre-entry that affect her own post-entry incentives. Perhaps the monopolist is engaged in a number of similar situations and will respond aggressively to entry in this market because she wishes to maintain her reputation for acting aggressively. On the other side, perhaps the monopolist cannot commit in advance to a production quantity if there is entry, and her incentives post-entry will lead her to be accommodating to the entry (in which case the story just told becomes rather dubious). All these are possible elaborations of the basic story that we can explore to develop our understanding of entry deterrence. And it is here that game theory comes into play; we use extensive form games as the language for building the different models, and we use solution techniques suggested by the theory to analyze those models.

To what end? In the next three sections, we will explore three particular classes of insights concerning dynamic competition that have been studied over the past twenty or so years. After discussing these classes of insights, we will be in a position to see the sorts of issues that can be treated successfully using game-theoretic techniques.

Incredible threats and incredible promises

A von Stackelberg story

The simplest way to elaborate the story of entry deterrence just told, which justifies the conjectures of the entrant more or less by fiat, is to suppose that the monopolist can and does commit to her level of output before the entrant has the opportunity to act. The extensive form game depicted in Figure 4.3 is suggested. The monopolist moves first, choosing her quantity.[2] The entrant observes this quantity choice and then decides whether to enter and, if so, what quantity to produce. Note carefully that each quantity choice by the monopolist leads to a single-node information set for the entrant; the entrant knows what quantity the monopolist has chosen when he chooses whether to enter and how much to produce; this is the point of this formulation. The figure is drawn out only for three quantities that the monopolist might choose and for three by the entrant, but the basic structure should be clear.

The payoffs and a detail of the formulation need a bit of explanation. In Figure 4.3, it is assumed that the market for this good extends for one period only. Hence if the monopolist chooses production quantity x and the entrant chooses not to enter, the payoffs, evaluated in units of net profit, are $(13 - x)x - x - 6.25$ for the monopolist and zero for the entrant; and if the monopolist chooses x and the entrant chooses y, their profits are $(13 - x - y)x - x - 6.25$ and $(13 - x - y)y - y - 6.25$, respectively. This formulation is not quite in accord with the verbal formulation of the problem in the story; in the story, there is a sense that the monopolist produces in a first period when she has the market all to herself, and then in a second period the entrant chooses whether to enter and, if so, at what level of production. If we think in terms of this alternative story, and we assume that the monopolist *must* produce in the

---

[2] In theory, the monopolist can choose from an infinite set of actions, as the quantity variable is perfectly divisible. If you find this bothersome, it will do to think that quantities must be in units no finer than hundredths.

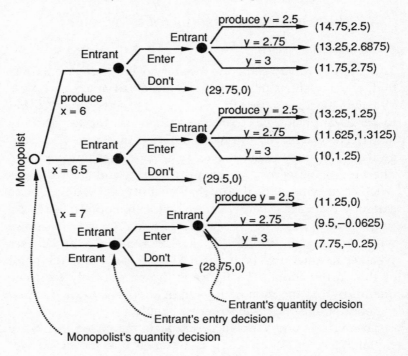

FIG. 4.3. The von Stackelberg game

second period the precise quantity that she produces in the first, then the extensive form game in Figure 4.3 would be the right model except for the payoffs. We would reinterpret the first choice (by the monopolist) as her single choice of a production quantity for each period. And then the monopolist's payoffs would be $[(13-x)x - x - 6.25] + [(13-x-y)x - x - 6.25]$ overall, where (a) the quantity in the first set of brackets gives the monopolist's profits in the first period, (b) the quantity in the second set of brackets gives her profits in the second period (where $y = 0$ is not ruled out, in case the entrant chooses not to enter), and (c) the fact that we sum the two levels of profits means that each unit of profit today is fully equivalent to a unit of profit tomorrow; if profits were discounted, we would modify the formula by including the discount factor.

Having posed the model in this fashion, we can proceed to analysis. We might try to solve the game using dominance arguments or by looking for a Nash equilibrium. Following the notions developed in Chapter 3, to do either we first recast this extensive form game as a strategic form game. But this quickly becomes extraordinarily tedious. Even if we imagine that the monopolist can choose only from among the three quantities shown in the figure and the entrant can choose from among the three quantities shown or not to enter, this means that the monopolist has three possible strategies and the entrant has sixty-four. Why sixty-four? If the monopolist can choose from among three possible quantities, then the entrant has three information sets. At each of these, the entrant has four possible actions, and so in the fashion of the 'menus of strategies' of Chapter 3, this gives the entrant $4 \times 4 \times 4 = 64$ strategies.

Rather than do this, we can argue directly from the extensive form.[3] Suppose the monopolist chooses to produce 6.5 units. The entrant knows this and now must pick a quantity. In Figure 4.4 we have reproduced the 'relevant' part of the game tree for the entrant, and we see that among the four choices we have left for the entrant, the best choice is to produce 2.75 units, for a net profit of 1.3125. Even if we had not restricted attention to these four possible responses by the entrant, it turns out that producing 2.75 is his best response to 6.5 produced by the monopolist. To see this note that the entrant's profits from

[3] (*a*) The direct argument that will be given is very closely connected both to notions of dominance and Nash equilibrium. This connection is sketched very briefly in Chapter 4; virtually any textbook on non-cooperative game theory will provide a more detailed explanation. (*b*) Recall from Chapter 3 the question, Is an extensive form game the 'same' as the strategic form game to which it corresponds? It should be clear from this example and others to follow that it is easier for us to think about and analyse some games that are presented in extensive form, even if there is a strategic form counterpart to the extensive form reasoning we use. This does not mean that there is any 'strategic difference' between a game in extensive form and the corresponding strategic form game, but it does suggest that there may be a perceptual or cognitive difference.

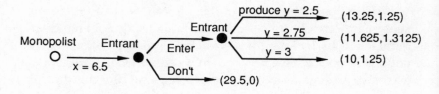

Fɪɢ. 4.4.   If the monopolist produces $x = 6.5$

producing y units are $(13 - 6.5 - y)y - y - 6.25$ if $y > 0$ and are zero if $y = 0$; maximize the quadratic function in y, and you will find that $y = 2.75$ is indeed the maximizing value of y and that it gives strictly positive profits.

In general, we can compute the optimal response of the entrant to every choice of quantity output x by the monopolist. If the monopolist produces x, the entrant seeks y to maximize $(13 - x - y)y - y - 6.25$; this is maximized at $y = (12 - x)/2$, at which point the entrant's net profits are $(12-x)^2/4 - 6.25$. (This is easy algebra as long as you know how to take the derivative of a quadratic.) Of course, the entrant has the option of not entering, which gives him profits of zero, and he will do this if his maximal net profits after entering are not positive. That is, the entrant chooses not to enter if $(12 - x)^2/4 - 6.25 \leq 0$, which is $x \geq 7$.[4] Summarizing, the entrant will enter and produce $(12 - x)/2$ if $x < 7$, and he will choose not to enter if $x \geq 7$.

Assuming the monopolist understands all this, the monopolist sees her own profit function as being

$$\pi(x) = \begin{cases} \left(13 - x - \frac{12-x}{2}\right) x - x - 6.25, & \text{if } x < 7, \text{ and} \\ (13 - x)x - x - 6.25, & \text{if } x \geq 7. \end{cases}$$

Let us explain: If $x < 7$, the entrant will enter, and the scale of entry is the term $(12 - x)/2$ in the first line; if $x \geq 7$, then the

---

[4] The second root of the quadratic equation $(12 - x)^2/4 = 6.25$, $x = 19$, is clearly extraneous. Also, we assume that the entrant chooses not to enter when he is indifferent between entry or not. There are good theoretical reasons for this assumption, but they are a bit hard to explain and are left to your further study of this sort of model.

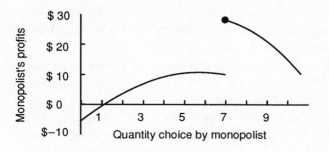

Fɪɢ. 4.5. The monopolist's profit function

entrant will not enter, hence the monopolist's profits are given by the second line. This profit function is graphed for you in Figure 4.5; note in particular the jump discontinuity at $x = 7$, and note that the monopolist's profits are maximized at $x = 7$, just as in the verbal story told previously.

An incredible threat

Suppose that just before the monopolist was to select her quantity in the game of Figure 4.3, the entrant made the following statement: 'Should you happen to choose a quantity x other than $x = 2$, I will respond with the quantity $y = 13 - x$, which will depress prices to zero and cause you to take losses. If, on the other hand, you choose the quantity $x = 2$, I will respond with $y = 5$.' How should the monopolist respond to this threat?

If she believes the threat, then there is no point in choosing any positive quantity other than $x = 2$, since to do so guarantees a loss even gross of her fixed costs. She should either choose to produce nothing or $x = 2$. If she chooses $x = 2$ (and the entrant chooses $y = 5$), then price will be $13 - 2 - 5 = 5$, and her profits will be $5 \times 2 - 2 - 6.25 = 1.75$, which is greater than zero. So her best response is to choose 2.

And the entrant's optimal response to a choice of 2 by the monopolist is indeed $y = 5$, which is obtained by maximizing

$(13 - 2 - y)y - y - 6.25$ in y. So if the monopolist finds this threat credible and responds optimally to it, the entrant will be happy to carry out his part of the bargain.

The problem is that the threat is hardly credible. One can easily imagine the monopolist grinning to herself as she chooses to produce 7 and then saying to the entrant, 'Be my guest. Choose $y = 6$.' The monopolist's smile will quickly fade if the entrant does indeed choose $y = 6$, but to do so causes the entrant to take a loss, and in the absence of other considerations (see the next two sections), it seems unlikely that the entrant would actually carry out the threat.[5]

The point is extremely simple. The analysis we performed in the previous subsection was based on the assumption that the entrant would respond to the monopolist's action in whatever manner was optimal for the entrant, given the monopolist's action. A threat to act otherwise is not credible, since the entrant must regard the monopolist's action as a *fait accompli*, which is precisely what it is.

Incredible promises

This is an example of an incredible threat. It is easy to think as well of promises that one party might wish to make but that the party in question will later have an incentive to break.

For example, an artist, having created a mould from which she can make a large number of castings, may wish to promise to buyers that she will cast only a small number, so that each casting will be valuable owing to its rarity. But having made such a promise and sold the small number at a price appropriate for the case where only a small number will be cast, the artist has the incentive to go back on her word and prepare

---

[5] You might worry that, while the entrant will not choose $y = 6$ if $x = 7$, he might choose to enter and produce $y = 2.5$, earning zero profits, just to spite the monopolist. If the entrant is motivated by spite, then to deter entry the monopolist must choose x sufficiently greater than 7 so that the cost to the entrant of entry at an optimal scale is just sufficient to offset his pleasure at spiting the monopolist.

more castings, to be sold at a lower price. Without some way of guaranteeing that she will not do this (such as a public breaking of the mould after only the small number has been cast and before they have been sold), buyers will regard her promise not to do so as incredible. Whatever her incentives are beforehand, after she has sold the small number her incentives will be to cast and sell more. Accordingly, buyers will not believe her promise in the absence of some credible guarantee and they will not pay the premium price appropriate if only a small number will be cast. (In a somewhat more general setting, this particular problem is known as the problem of the durable-goods monopolist.)

Imagine a government that controls the money supply of a particular country. For political reasons, the government wishes to run the economy so as to maximize its growth. Assume that anticipated inflation is deleterious to growth, but that unanticipated inflation (fuelled by an unanticipated increase in the money supply) enhances growth. Taking the first effect into account, the government might wish to promise not to inflate the money supply at more than a given rate. But *ex post* the government will wish to pump money into the economy, to take advantage of the second effect. The horns of this dilemma are that the government's *ex ante* promise not to inflate is incredible insofar as the general public recognize that, *ex post*, it will be in both the government's best interests and its power to inflate the money supply. Hence the public will anticipate high inflation, which depresses growth.

To take an example from public finance, consider the development of a natural resource, say an oil-field. In order to promote development, the government wishes to promise developers that exorbitant taxes will not be imposed on revenues earned on oil pumped out of the ground. But *ex post*, as development of the oil-field is completed and the pumping begins, there are changes in the government's incentives concerning what tax rate to levy. If the government cannot guarantee what taxes will later be imposed (guarantees made more difficult by the fact that the current government may be replaced by a different government, formed by another party), the *ex ante* preferable promise of a perpetually low tax levy is incredible; firms developing the oil-field should do so anticipating that a high tax rate will be imposed later.

A fourth example concerns vertical integration of two firms. Imagine a firm that purchases a particular product from a second firm. The first firm decides that for various reasons it is sensible to buy out the second, making the previously market-based transaction an internal transaction. But management in the first firm is worried that management in the second firm, who provide expertise and technical know-how, will leave. Accordingly, management in the first firm promises that management in the second firm will retain its autonomy and will be rewarded just as if the two were separate firms. Insofar as integration means that the first firm's managers will have the legal ability to abrogate such promises, either directly or indirectly through its control of accounts within the firm, and insofar as the *ex post* incentives of the first firm will be to abrogate (or at least shade) those promises, then the second firm's managers will find the promise not to do so incredible.

In labour economics, management in a piece-rate workplace may wish to promise that any resetting of the rate will not take advantage of information gained from the performance of workers but will be based only on changes in technological factors. In international trade, one government may promise a second to liberalize trade practices next year (or in five years, or 'as soon as circumstances warrant') in exchange for concessions given now. There is a long list of threats and promises that lack credibility in dynamic settings because *ex ante* incentives to make the promises or threats do not match the *ex post* incentives to carry them out. A notable success of non-cooperative game theory is that it provides both the means of formalizing the context of such promises and threats (in an extensive form game) and then analysing the credibility of the promises and threats.

I hasten to add that the specific analysis given previously— the entry-deterring monopolist who fixes his quantity before the entrant has an opportunity to act and in view of the entrant —cannot be counted as the product of game-theoretic think- ing; it (essentially) was produced by von Stackelberg (1934) long before the modern development of game theory. In this instance game theory may be helpful in explaining the 'rules of the game' (i.e. in producing Figure 4.3), but that is about all.

Non-cooperative game theory has added quite a lot to the general program of analysing credibility, however. In the first place, the ability to draw the pictures gives us an ability to frame the issues and raises consciousness about their import- ance. Moreover, non-cooperative game theory has provided economists with a unified language for phenomena related to credibility, which permits cross-contextual comparisons. And not all situations are as straightforward as the game in Figure 4.3; non-cooperative game theory has supplied tools that can transfer the simple ideas we applied previously to increasingly complex settings. For the remainder of this section, we will indicate how this last programme has been carried out in the context of entry deterrence.

Games of complete and perfect information and backward induction

To begin, we formalize and generalize the exact procedure that we applied in the analysis of Figure 4.3, when we ruled out the incredible threat posed by the entrant to respond to $x \neq 2$ with $y = 13 - x$. The game in Figure 4.3 has a very nice prop- erty, namely that every node in the game tree is, by itself, an information set. The game depicted in Figure 3.3 (reproduced here as Figure 4.6) is another example of such a game, and the game of chess is a third example. Such games are called *games of complete and perfect information*.

The common-sense procedure we used previously applies

very nicely and directly to games of complete and perfect information.[6] Go to the end of the tree and choose a node, all of whose successors are final payoffs. What will happen at that node is a matter of common sense; the player whose choice it is will choose to end the game in the fashion most advantageous to himself. Once you know what will happen at all such 'almost-terminal' nodes, you can discover what will happen at 'almost-almost-terminal' nodes, or nodes all of whose successors are either payoffs or almost-terminal. And so on, all the way back to the start of the game.

If you aren't clear on how this works, consider the game in Figure 4.6. There are two almost-terminal nodes, viz. the node belonging to B in the lower-left of the figure and the node belonging to C all the way to the right. At the first of these, B is clearly going to choose Y (and get 3 rather than 2), and at the second, C is going to choose w'. Now consider the node belonging to B just to the left of C's node on the right. B has a choice between X' and Y'. If B chooses Y', he gets a payoff of 3, while choosing X' gives the move to C, which we just said will lead to a choice of w', yielding 2 for B. Hence B will choose Y'. Hence one move prior to this, A will choose y', and hence one move prior to this, C will choose u. Now we are back to the initial move of the game. A can choose x and get 3, or y (giving the move to B, who will choose Y) which results in 2 for A, or z (giving the move to C, who will choose u) resulting in 1 for A. Hence A chooses x.

In Figure 4.6 this procedure takes more steps than in the game of Figure 4.3, but the procedure is basically the same. Indeed, we can apply the same procedure to the game of chess. Chess is a game of complete and perfect information. It has a finite game tree. So all we need to do is to draw out the

---

[6] To be precise, it applies very nicely to games of complete and perfect information with the further property that there is a finite upper bound to the number of nodes along any path from the start through to the end of the game, and it applies nicely to such games except for one caveat to be given subsequently.

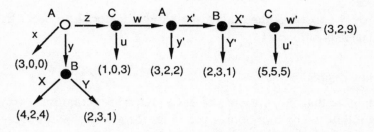

Fig. 4.6. A game of complete and perfect information

game tree for chess and, starting from the back (one move from the end of the game), work out the best move for whoever has the move, and then, node by node, work our way to the start of the game, finding the 'optimal strategy' for both sides. Instant grandmastership!, except that in the case of chess, the procedure is completely impractical; the game tree is too big.

But when the game tree isn't too big, this common-sense procedure, known as *backward induction*, seems a very nice way to 'solve' games of complete and perfect information.

(There is one complication that we don't see in the game in Figure 4.6 but that appears in the game in Figure 4.3. In 4.6, as we moved back through the tree, we never came to a point at which a given player was indifferent between two or more of the actions he could take. Hence there was always a clear choice to be made. Think back to the game in Figure 4.3 and the node at which the entrant must choose what to do if the monopolist chose output level $x^0 = 7$. If the entrant chooses to enter, the best he can do is to respond with $y = 2.5$, which gives him net profits of zero. Alternatively, he can choose to stay out, which gives him net profits of zero. He doesn't care which he does, but the monopolist is intensely interested; she far prefers that the entrant stays out. When there are ties as one uses backward induction, it can matter how the ties are broken; this leads to complications that I will ignore in this book by resolving ties in a convenient fashion. This convenience often has some theoretical justification; you should consult something resembling a textbook to see what that justfication is.)

More complex variations

Imagine that the story of the entry-deterring monopolist runs as follows. The monopolist produces for a while, without any possibility of entry occurring. During this period, the monopolist sets a price p, and demand for her product is given by $x = 13 - p$. The monopolist's fixed costs for this period are 6.25. Then there is a second period of time. At the start of this second period of time, the entrant must decide whether to enter. If the entrant does not enter, then the monopolist remains a monopolist and can (if she wishes) reset the price she charges or, equivalently, the quantity she sells. If the entrant does enter, then the monopolist and the entrant compete on equal footing. Demand during this second period is given by the demand function $x = 13 - p$, and the monopolist and the entrant (if he enters) both pay fixed costs of 6.25.

The key that distinguishes this second formulation from the first is that if the entrant does enter, then in the second period the monopolist and entrant are on an equal footing. Nothing the monopolist did in the first period affects the capabilities of either party or the opportunities each has.

In order to apply game theory to analyse this situation we need a model of duopolistic competition. We will employ the game-theoretic model associated with Cournot equilibrium in which the two duopolists simultaneously and independently choose production quantities and price is set in the market so demand is just equal to the total supply. If this is our model of duopolistic competition, then Figure 4.7 shows the overall extensive form game model of the situation that we seek to analyse. (Once again, we only work our way through a few of the branches in the tree.) First the monopolist chooses her price–quantity pair in the first period. For example, one choice she could make, which is shown, is to sell six units for a price

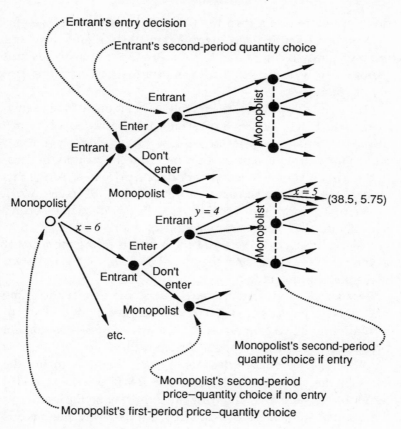

Fig. 4.7. A formulation of the entry-deterring monopolist story
In this extensive form game, the monopolist sets her quantity
in each of two periods, and her first-period choice has no effect
on what is feasible or on the profits in the second period.

of 7. The entrant observes this and decides whether to enter.
Note that the entrant's presumed ability to observe the monopolist's first period choice of price–quantity is depicted by the
lack of an information set for the entrant at this point. If the
entrant chooses not to enter, then the monopolist again chooses
a price–quantity pair for the second period. But if the entrant

does choose to enter, then the Cournot game ensues; the entrant picks a quantity, and the monopolist simultaneously and independently picks a quantity. As usual, 'simultaneously and independently' is modelled with an information set, in this case for the monopolist.

To see how payoffs are made, follow in Figure 4.7 the branch along which the sequence of events is: the monopolist produces six units for a price of 7 in the first period; the entrant enters; the entrant produces four units and the monopolist produces five units in the second period for a price of 4. Along this branch, the monopolist earns profits of $7 \times 6 - 6 - 6.25 = 29.75$ in the first period and $4 \times 5 - 5 - 6.25 = 8.75$ in the second period, and the entrant earns profits of $4 \times 4 - 4 - 6.25 = 5.75$ in the second period. We sum up the profits of the monopolist in the two periods, giving payoffs of 38.5 and 5.75 to the two players along this branch.[7]

Because of the information set out at the end of the game tree, this is no longer a game of complete and perfect information and we cannot use backward induction to make a prediction what will happen. Instead, the following logic is employed: *If* the entrant decides to enter, then the 'rest of the game' is the piece shown in Figure 4.8. I will use the standard terminology hereafter, referring to this 'rest of the game' as the *Cournot subgame*. Note carefully in this figure the payoffs to the monopolist; at the end of the branch where the monopolist chooses output level x and the entrant chooses y, the monopolist's payoff is written $\pi_1 + (13 - x - y)x - x - 6.25$. The term π_1 is the monopolist's profits carried forward from the first period; the same π_1 applies to all the branches following a single first-period decision by the monopolist, although the value of π_1 changes with that first-period decision.

---

[7] A standard procedure in such models is to assume that the monopolist discounts future profits at some given discount rate; adding this feature won't change the basic conclusions we will come to and would add needless complications, so we assume instead that the monopolist wishes to maximize the undiscounted sum of her profits.

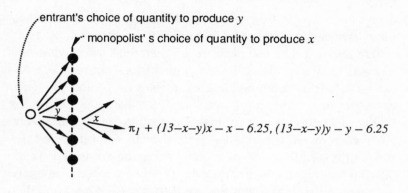

entrant's choice of quantity to produce y

monopolist' s choice of quantity to produce x

$\pi_1 + (13-x-y)x - x - 6.25, (13-x-y)y - y - 6.25$

FIG. 4.8.   The Cournot subgame, if the entrant enters

Each Cournot subgame is an extensive form game in its own right, and the key step in the analysis of the entire situation is the assumption that *whatever transpires in the first period, if the entrant enters the two firms will select Nash equilibrium strategies for the subgame.*

The logic behind this assumption is an extension of the logic of no incredible threats or promises. The entrant might threaten the monopolist that he will do this or that if the mono- polist produces at some level or another in the first period, but such a threat or promise is only credible if, after the fact, it is in the entrant's own interests to carry it out. And, in this formu- lation, a similar constraint applies to the monopolist. To scare the entrant into staying out, she might threaten that she will produce some very large quantity in the second period if he deigns to enter. But once (and if) the entrant has entered, the monopolist is faced with a *fait accompli,* and she is assumed to carry out actions that are in her own best interests given the circumstances.

It is not trivial to jump from these arguments—arguments for what might be called *ex post rationality*—to an assertion that the two parties will play some Nash equilibrium. This leap assumes that in each circumstance that could conceivably occur, the two parties see 'self-evident ways to continue to play', which (then) must be in equilibrium with each other.

But laying aside the problem of that final leap, when we proceed to analysis the following conclusions are reached: No matter what the monopolist does in the first period, the sub-game (if the entrant enters) has a unique Nash equilibrium. In that equilibrium the entrant makes a positive profit. Moreover, the profits of the monopolist derived from second-period activity in the equilibrium of a given subgame are independent of what she does in the first period. (These assertions are likely not to be evident, and you will need to consult a textbook that discusses Cournot equilibrium from the point of view of non-cooperative game theory to see them proved.) Hence (now applying backward induction) whatever the monopolist does in the first period, the entrant will enter. And thus the monopolist's best course of action from the start of the game is to make as large profits as possible in the first period; nothing done by her in the first period is going to change the entrant's actions or her second-period profits. In this formulation, the entry-deterrence story falls apart.

The intuition behind this is really quite simple. In this formulation, there is no reason for the entrant to pay attention at all to the first-period price–quantity decision of the monopolist. It changes nothing about the conditions of competition between the two if the entrant chooses to enter. Once the entrant enters, the two are on completely equal footing.[8]

In the context of this two-period formulation, we are going to be able to resurrect entry deterrence only if we think of some way in which the first-period decisions of the monopolist affects the terms of competition in the second period. Several possibilities can be suggested: The monopolist in the first period chooses her production technology; she is able to

---

[8] For those readers who know a bit about this sort of analysis, let me add the caveat (which you have probably already added yourself) that the argument depends as well on the uniqueness of the subgame equilibrium.

trade off lower marginal costs for higher fixed costs. Insofar as having a lower marginal cost and a higher fixed cost makes her more aggressive in any second-period competition, she may pick a technology with suboptimally high fixed costs in order to forestall entry. She may be able to delegate decision-making authority to some manager whose incentives are to act aggressively. If the good is one in which there is an element of consumer loyalty, she may sell more at a lower price in the first period to build up her base of loyal consumers sufficiently to forestall entry. She may take actions in the first period that *create* consumer loyalty. If the good is one that is somewhat durable, she may sell more in early periods in order to dampen demand later (and forestall entry). If the good is one that can be differentiated (sold in a number of different varieties), and if the cost of presenting a number of varieties is somewhat sunk (so that if she offers, say, six varieties in the first period, it is relatively cheaper for her to continue with six varieties in the second period), then she may proliferate varieties beyond the point that would be optimal if she didn't have to worry about entry, so as to leave no 'holes' in the market through which the entrant can profitably enter.

All these elaborations on the basic story can be fleshed out and examined using the techniques of non-cooperative game theory; many of them, and many others, have been examined in the literature. They share the characteristic that the monopolist in the first period undertakes an action that changes the nature of the subgame if the entrant enters, a change that is sufficient to forestall entry. (Or, even if entry is not forestalled, the monopolist may find it in her interest to prepare for entry so that, post-entry, she has an advantage in the resulting duopolistic competition.)

Rather than pursue any of these variations (this is not, after all, a book in industrial organization), let me propose a further elaboration of the story that takes us a step further into complexity.

A more subtle way in which the first-period actions of the monopolist might affect competition in the second period is if the entrant is uncertain about characteristics of the monopolist and regards the monopolist's early actions as a signal of those characteristics. For example, suppose the entrant is uncertain about the cost structure of the monopolist. Suppose that if the monopolist has particularly high unit costs, then in a duopoly the entrant will be able to prosper. But if the monopolist has low unit costs, then duopolistic competition will be too severe for the entrant. If the entrant must pay a sunk cost for entering the industry, he will guess at the outset what the costs of the monopolist might be. A classic monopoly firm typically charges a higher price the higher are its marginal costs.[9] Hence the entrant might perceive a relatively low price and high quantity in the pre-entry period as evidence that the monopolist has low marginal costs and hence as a signal to stay out. And the monopolist, thinking that the entrant will think this way, may decide to charge a price that is lower than the short-run optimum in the pre-entry period to convince the entrant that she has low marginal costs.

To flesh this story out formally takes us very far beyond the tools we have available here. The bottom line of the analyses that appear in the literature is that this story of entry deterrence may work, but it is not quite as simple as the glib description just given. While we cannot give the analysis,[10] we can at least indicate why the game-theoretic tools that are required are (at least) a level more complex than those so far discussed.

Imagine a very simple version of the story in which the monopolist's unit costs are either high or low. The monopolist knows her own unit costs, but the entrant is uncertain. Otherwise the structure of the 'game' between the two is much the

---

[9] Draw the picture of a classic monopolist who equates marginal cost and marginal revenue and assume that both demand and marginal revenue are decreasing functions of quantity. What then is the effect of an upward shift in the marginal cost curve?

[10] See the end of this chapter for some suggested readings.

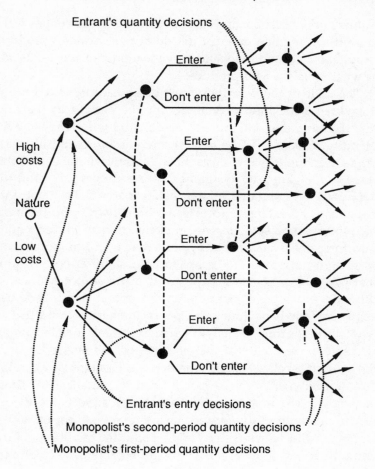

Entrant's quantity decisions

Enter

Don't enter

Enter

Don't enter

Enter

Don't enter

Enter

Don't enter

High
costs

Nature

Low
costs

Entrant's entry decisions

Monopolist's second-period quantity decisions

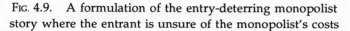
Monopolist's first-period quantity decisions

FIG. 4.9. A formulation of the entry-deterring monopolist
story where the entrant is unsure of the monopolist's costs

same as in Figure 4.7. Then the extensive form schematically
given in Figure 4.9 is suggested. Notice that the first move
in this game is a move by Nature, which determines the unit
costs of the monopolist. Then, depending on what Nature
chooses, the monopolist chooses her first period price–quantity
pair. The entrant observes this price–quantity choice *but not*

Nature's initial selection and must decide whether to enter. If the entrant enters, then he must decide at what scale to produce; simultaneously and independently the monopolist must decide at what scale she will produce.

The italic in the description just given is represented in the diagram by the information sets for the entrant at the time that the entry decision must be made. And when choosing his quantity to output (if he enters), the entrant is still unaware of the monopolist's unit costs; note the information sets at this stage of the game. In this game, there is no point beyond the very start of the game in which 'the part of the game that remains' is an extensive form game in its own right; the complex information sets present problems. (Since I haven't quite said formally what I mean by 'an extensive form game in its own right', my assertion is apt to be obscure at best. I hope that you will be able at least to see that these information sets present a problem.) We need a formalism that allows us to talk about credible behaviour in pieces of the game. In particular, when it comes time to analyse whether the entrant's behaviour is credible, credibility of the entrant's behaviour depends on the entrant's beliefs about whether the monopolist has high or low unit costs. Thus we will need a formalism for those beliefs. Recent innovations in non-cooperative game theory provide us with those formalisms and even present us with a language with which we can argue about the credibility of the entrant's beliefs as a precursor to arguing about the credibility of the entrant's behaviour, based on those beliefs.

To push any further without subjecting you to a full course in the subject is impossible. But I hope you take the point: To study stories such as the story of the entry-deterring monopolist, we must study dynamic interactions between the parties involved. Answers we get will turn on what one party believes the second will do in response to actions by the first; and so we require a technology for exploring the credibility of those conjectured responses. Beginning with the simple and straightforward story of von Stackelberg, non-cooperative game theory gives us the tools for studying this problem in increasingly complex settings, so that we can probe and extend our initial common-sense intuition. As we do this in increasingly more complex situations, intuition although aided by the analysis may begin to fail, at which point it becomes time to stop. But even in the more complex settings in which the theory does not give definitive answers, such as in the final formulation where the entrant is uncertain *ex ante* what are the marginal costs of the monopolist, the theory suggests where we must either apply our intuition or extend our empirically generated knowledge in order to proceed.

Credible threats and promises: Co-operation and reputation

So far we have concentrated on techniques for concluding that certain threats or promises are not credible. But in many cases, promises (and threats) are credible. As noted in the context of entry deterrence, this may be because the promise is accompanied by actions which make keeping the promise optimal for the promise-maker *ex post*. But a second major success of game theory has been that it suggests ways in which promises and threats are credible, because the promise-maker stakes his or her reputation on fulfilment.

Simon's (1951) analysis of the employment relationship is a very early (pre-game-theoretic) application of this notion. Simon considers a case in which employment of party B by party A amounts to the following sort of open-ended contract: For wages that are set at the start, party B agrees to accept otherwise unspecified direction by party A. The contract is kept open-ended, according to Simon, because it is sometimes too difficult to specify a priori just what tasks A will wish B to complete; as contingencies arise, A sees what needs doing and so directs B. But then B should worry whether A will ask for too much or for too onerous a set of tasks. Of course, B retains the right to quit (in most societies), but insofar as leaving one job for another entails a period of unemployment, loss of the companionship of fellow workers, perhaps the costs of retraining or relocation, and so on, B's bargaining position with A once B has begun to work for A is eroded relative to his position *ex ante*. Why doesn't A exploit this *ex post*? Why doesn't B worry that A will exploit this?

One possibility is that B is aware of A's *ex post* incentives and, even factoring those incentives into account, the job is better for B than his next best alternative. A second possibility is that just as B becomes hostage to A's whims (in that B will incur costs by invoking his right to resign), so A becomes hostage to B; B may learn things about A's business and so become relatively irreplaceable or, at least, replaceable only at the cost of training a new worker. This may strenghten B's hand in any *ex post* 'negotiations' over whether B will perform certain tasks or quit, and it may restrain A from asking for too much.

Simon suggests as well a third possibility. Party A may promise either implicitly or explicitly that the job will not entail overly onerous or distasteful duties. And this may be a credible promise because if A should violate it, then A will become known *generally* as a bad employer. If A's reputation in the community is bad, then she may be unable to find another worker if B does quit. It is not the loss of B's work that threatens and restrains her; rather she is restrained by the possible loss of her reputation, which in turn would mean an inability to hire others or the need to pay premium wages to attract others.

A stylized model of reputation

Non-cooperative game theory has given us an excellent formal account of this intuitive notion and of notions related to it, and

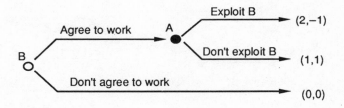

Fig. 4.10. A stylized example of the employment relationship

it has given us the ability to explore important qualifications to and aspects of the basic story. To take a very stylized example, suppose the situation is as depicted in Figure 4.10. Worker B must decide whether or not to accept employment with A, and then A must decide whether or not to exploit B. B is willing to work for A if and only if A does not exploit him. But if B does agree to work for A, then it is in A's interests to exploit him. Hence by backward induction, A will exploit B if B agrees to work for A, and B declines employment. Of course, if A could credibly commit not to exploit B, then B would agree to employment, and A and B would both be better off.

Now imagine that A plays this game against not a single B but a sequence of Bs; first with B1, then with B2, and so on. Each Bn is interested only in his payoff from his interaction with A. For A, on the other hand, an outcome is an infinite sequence of results—what happens with B1, what happens with B2, and so on—and if u_n is A's payoff in her encounter with Bn, then A evaluates the infinite sequence of payoffs $u_1, u_2, \ldots$ according to the discounted sum of this sequence, or $u_1 + au_2 + a^2u_3 + a^3u_4 + \ldots$ for some number a strictly between zero and one. And, crucially, employee Bn, when deciding whether to take employment with A, is aware of A's past history of treatment of workers.

Drawing the complete game tree for this game is impossible, but a start on the picture is given in Figure 4.11. This is only a start because it only carries us through the first two

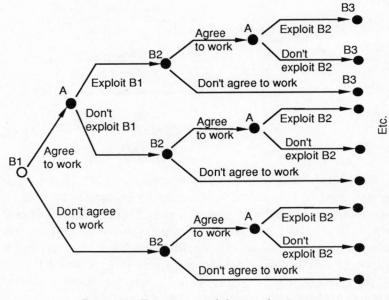

Fig. 4.11.   Two stages of the employment
game with infinitely many stages

'stages'. This is a game with an infinite number of players,
A and all the Bn, and the complete game tree has paths of
infinite length. Rather than thinking of a path through the tree
culminating in a vector of payoffs after some finite sequence of
actions, you must think of an infinite sequence of actions and
a function which assigns for each player in the game and for
every possible infinite sequence of actions a payoff. You can
think of the discounted and summed payoffs for A as something like compound interest; imagine that the payoffs are in
monetary units, each stage takes a month, and $a = 1/(1 + i)$
where i is the rate of interest per month on money. For the
sake of concreteness, we hereafter assume that $a = .9$.

At the risk of repeating the obvious, note that B2 has three
information sets and B3 has nine. B2 knows whether B1 took
a job with A or not and, if so, how B1 was treated by A, and
B3 knows all this about B2 and B1.

We couldn't conceivably depict the strategic form game that is related to this extensive form game, since the number of players is infinite (and with just three players we had to speak of multi-storey parking-lots), the number of strategies for Bn is $2^{3^{n-1}}$, and the number of strategies for A is infinite. And while this is a game of complete and perfect information (all the information sets are single nodes), we can't apply backward induction because the tree isn't finite; there is no last node with which to begin.

None the less, we can describe Nash equilibria of this game. And among the Nash equilibria is the following: Employer A carries a reputation for how she treats her workers. She begins with a good reputation, which she keeps as long as she never exploits a worker. If she ever exploits a worker, she gets a bad reputation, which she can never shed. Note that A's reputation based on her treatment of the workers B1 through $B(n-1)$ is known to Bn. Then Bn will agree to work for A if and only if she has a good reputation. And A exploits worker Bn if and only if she already has a bad reputation.

Why is this a Nash equilibrium? In stage n, if A has never exploited a worker in the past, her strategy calls for her not to exploit Bn, and then the best response of Bn is certainly to take employment. On the other hand, if she ever exploited a worker previously, then Bn can expect to be exploited, and he is better off turning down her offer. If A has ever exploited a worker in the past, then she won't get the opportunity to exploit worker Bn, but even if by some fluke she was given this opportunity, she might as well exploit him because no one will ever trust her again, no matter how she behaves.[11] The

---

[11] For readers who know pieces of game theory that I haven't formally introduced here, note that this part of the argument is important in a demonstration that these strategies form a subgame perfect equilibrium. For readers who don't know much more than is covered here, what this means is that this particular equilibrium is not based on any incredible threats or promises. No matter what happens, players will always wish to carry out their parts of the equilibrium given that other players subsequently conform.

key question is, Why doesn't A exploit Bn if she has never previously exploited an earlier worker? The answer is that her reputation is, by construction, worth more to her than any short-run benefits she can obtain by exploiting a single worker. As long as she maintains her reputation, she gets a payoff of 1 each period, giving her a discounted present value of payoffs equal to 10.[12] If she exploits a worker, she gets 2 in that period, but no worker ever again agrees to work for her, so her total payoff is 2. Since 10 > 2, she clearly prefers to keep her reputation intact.

Of course, the outcome of these strategies is that B1 takes employment with A, A does not exploit him, hence B2 takes employment with A and is not exploited, and so on.

This construction is not without its problems. Let me briefly mention a few of the more obvious:

(1) The key to the argument is that A always has a reputation to protect and the value of that reputation always exceeds the short-run gains she could obtain from sullying her reputation. It is crucial, then, that this game should have an infinite tree. If at the start of any round the two players involved know that this is the last, then A will certainly exploit this B given the opportunity—no point in protecting a reputation if there are no further opportunities to use that reputation—so B will certainly not give A the opportunity. So if at the start of some round the players involved know that either this or the next will be the last, then A will exploit this B. Either this is the last round—no point in keeping a good reputation—or next round will certainly be the last—the next B will not enter into employment no matter what A's reputation is, so again a good reputation isn't worth anything. Hence this B will not enter into employment with A. And so on. If the game tree is finite, we can apply backward induction, and applying backward induction causes the entire reputation construction to collapse.

[12] In case you missed it, the discount factor a is .9, so $\sum_{k=1}^{\infty} a^{k-1} 1 = 1/(1-a) = 10$.

(2) The equilibrium we have given is but one equilibrium. Another is that A exploits every B that enters into employment, and every B refuses to work for A. And there are many others besides. By applying the logic of Nash equilibrium, we implicitly maintain that there is a clear-cut way to behave in this situation. But we haven't offered any reason to believe in our 'reputation' equilibrium over others, and while we might be able to make a good case for it in this simple, stylized example, the case would turn crucially on the utter simplicity of this example; it would not translate well at all to more complex formulations, where choices are not so simply dichotomous.

(3) Our construction depends crucially on each Bn seeing precisely how A has behaved previously. But one can easily imagine more complex settings and formulations in which current employees have only a rough idea how A behaved in the past; a previous worker may claim to have been exploited, but this worker may have been a malcontent; A may not have realized when assigning a particular task how onerous it would become; and so on. We can wonder whether our construction has any chance of surviving realistic features such as these.

(4) Imagine that A exploits B1. The strategies in the reputation equilibrium call for no Bn ever to work for A. But A can then go to the Bs and say, 'I made a terrible mistake in exploiting B1, and I'm truly sorry about it. Please forgive and forget. And notice that it is in everyone's interests to forgive and forget, since as things stand I get zero payoff for the rest of time, but so do all you Bs.' Of course, if the Bs are very forgiving and let A get away with it this time, then B2 may find himself the next target of exploitation, followed by hearts and flowers for B3, B4, and so on. But at the same time, in more realistic applications, one has to worry that, when and if it comes time for a player to pay for any transgressions, all will see it in their *joint* best interest to forgive and forget. And anticipating that this will in fact happen, players will not believe in the power of reputation, hence reputation constructions will fail.

(5) In any realistic study of employment, the problem is much more complex than this. There are employees, foremen, and bosses of many different types. Does reputation attach to the business—'Gigantic Motors exploits its workers'—or to the shop—'Workers in the fender shop are exploited'—or to the individual foreman—'Don't work for Joan Jones'? Since reputation requires permanence to work, some element of the first two must be at work. But then what are the foreman's interests in all this? How is it that he or she is motivated to protect adequately the good name of the firm?

(6) Another realistic feature missed by this model is that the world is not so simply stationary as in this repeated game model. Time changes the jobs required, workers are not all the same, and so on. How does reputation work in such cases? On what is it based?

The first problem cited can be dealt with using the tools discussed in the next section. Problem (3) is an important avenue for developing and refining our intuition of reputation; imperfect observability and its impact on reputation have been an important area of research using game-theoretic techniques. Problem (4) has also been the subject of attention in the recent literature; it is another place in which the technology of game theory allows us to move a step or two beyond simple common sense. Problem (5) is characteristic of elaborations that take this simple model and use it to study issues of organization design. There is beginning to be some work on problem (6), but this remains a challenging elaboration. Problem (2) remains an enormous problem, and it will play a major part in Chapters 5 and 6.

Multilateral co-operation and the folk theorem

In the story of the previous subsection we imagined one long-lived employer who carried a reputation among a community of potential employees. Similar constructions are feasible and, in fact, somewhat enhanced when we think about a group of individuals, all of whom are engaged in repeated interaction with each other.

A classic example of this is implicit collusion in an oligopoly. For simplicity, we will consider the case of a duopoly. Two firms produce the same good for a market. They produce to-day, and tomorrow, and as far as either is able to see. Demand in each period is given by some downward-sloping demand curve. Each firm has a very simple production technology with zero fixed costs and constant marginal cost of c. They compete by posting prices; each takes out an advertisement in the morning newspaper saying at what price they are willing to sell their wares, and then customers queue up for the product. The entire queue goes to whomever named the lower price that day; in the event of a tie, demand is divided by some known rule. For the sake of concreteness we assume that when they name the same price half of the demand goes to each. The two duopolists are not capacity constrained; they can produce to demand.

I don't know if the institution exists outside of the United States, but in the United States camera and electronic equipment discounters operate in something like this fashion; if you have ever seen a Sunday edition of the *New York Times*, you cannot have failed to notice advertisements for 47th Street Photo and the like. Of course, the assumption of zero fixed costs is probably wrong in this example, but this assumption isn't really needed for the final conclusions (about implicit collusion) that we come to.[13]

If we imagine the two oligopolists competing once and once only, then the only possible equilibrium between them is for each to name a price of c (their marginal cost), hence each makes zero profits. The precise argument that this is so needn't concern us (and if you are overly concerned with the zero-profit outcome, there are simple variations in which each makes

---

[13] The case of 47th Street Photo and its competitors is different from the story given here on other, much more important grounds; see following.

a small positive profit, although the profit must be small); the intuition is that if either was making a substantial profit, it would pay the other to undercut the first's price by a little bit and take away all the business of the first.[14]

But if we imagine the two competing repeatedly, where each discounts the stream of profits received in the manner of employer A in the previous subsection, then we can come to a very different conclusion. Let p^* denote the price that a monopolist would charge in this industry. Suppose that each duopolist adopts the following strategy. Charge p^* as long as the other firm has charged p^* in the past, but if ever the other firm has charged any price other than p^*, charge c forever after. Putting it a bit more colourfully, each firm carries a reputation with the other for 'orderly, restrained, and polite' competition as long as it has charged p^*, a reputation that is replaced by a reputation for 'cut-throat' competition if ever another price is charged. Then the strategy above can be rewritten as, Compete in restrained fashion as long as your rival has a reputation for restrained competition, and slash prices otherwise.

As long as each firm puts substantial weight on its future profits, it is a Nash equilibrium for each to play the strategies given above.[15] Either can double its immediate profits by cutting price a bit in a single period, but this will trigger price-slashing by one's rival, and subsequent profits will be zero.

This is, of course, nothing more than a highly stylized version of the common-sense notion of a cartel. Each firm abides by the discipline of the cartel because each does better than if cut-throat competition ensues. Several remarks are in order:

(1) Note that in this case it is essential that both firms have a

---

[14] For the more sophisticated reader, this is just the classic Bertrand equilibrium.

[15] These strategies are not quite subgame perfect. But then if you know the terminology required to make sense of my remark, you probably already knew that and how to fix it.

stake in the future; if, as in the story of the employment relationship, we had one firm facing a sequence of opponents,
those opponents would undercut the one firm every chance
they got, so (by a slightly complex form of the usual Bertrand
argument) the one long-lived firm would necessarily charge
price c in each period as would all its rivals. (If you know
about such things, think about this in the setting of Cournot
competition or Bertrand competition with differentiated prod-
ucts. Then as long as the future matters enough, a single long-
lived firm can act like a von Stackelberg leader in each period.)
The point is that when both sides have a reputation to protect,
we can sometimes get 'co-operation' in cases where we cannot
with only one-sided reputation constructions.

(2) As in the stylized model of employment, there are many
equilibria possible in this case. In fact, a famous result in game
theory, known as the *folk theorem*, shows that any payoffs for
the two firms that give each more than zero (more than the
worst payoff the other can inflict upon them) and sum to less
than monopoly profits (per period) can be sustained in an equilibrium, if the future is weighted heavily enough by each. As
before, the problem of multiple equilibria is one of the weaknesses of this result, to be discussed in the next two chapters.

The fact that this result is called the folk theorem should
not go without comment. It is called that because the result is
held to be fairly obvious common sense. Formal statements
and proofs can be a bit complex, but the intuition is clear
and obvious; 'co-operation' of a given sort is held together
by a threat to punish those who transgress. As long as the future looms large relative to the present time, such threatened
punishments are effective.[16] And because this result is fairly
obvious common sense—something that game theorists (and,

---

[16] Are these threats credible? Part of the complexity of mathematical treatments of the result comes from working hard to show that the threats can be
made credible, where the standard sort of construction has the threat to punish
a transgressor made credible by a threat to punish anyone who doesn't punish
the transgressor, and so on.

certainly, most economists) feel they 'knew' all along—it is part of the folk wisdom of the subject; no one is brash enough to claim authorship of the general idea.

(3) Not only are there many equilibrium outcomes in this problem, but also many outcomes can be sustained by many different forms of punishment. For example, to sustain the outcome where each firm charges p^* in each period, it might do for each to threaten the other to charge c for five periods or ten following any defection. That is, we can think of equilibria in which punishments have the appearance of a 'price war'.

(4) As in the previous subsection, the construction can fall apart utterly, unravelling from the back, if there is a finite horizon. As before, elaborations along the lines discussed in the next section can come to the rescue (at least to some extent).

(5) As in the previous subsection, the inability to observe what the other party is doing and the possibility of forgiveness (or otherwise renegotiating the collusive scheme in the middle) present both problems for and areas of fruitful elaboration and extension of the basic scheme.

(6) An important real-life complication missing in this stylized example is that the results of competition today often affect the conditions of competition tomorrow. In our stylized example, grabbing part of the market today doesn't give to either firm any advantage tomorrow. But things become more complex and interesting when customers exhibit loyalties, when there is learning-by-doing, and so on. Again we have variations on the basic theme that have been fruitful areas of research.

(7) The case of 47th Street Photo and other electronics discounters does not fit the structure of the stylized example very well because in discount retailing, entry is fairly easy. The basic story of implicit collusion works best when there are substantial barriers to entry, so an oligopoly will stay an oligopoly, even if participants are highly profitable. You may find interesting both the case of electrical equipment (Sultan 1975,

Porter 1983)—a cartel that worked—and the case of airframe manufacture (Newhouse 1982)—a potential cartel that doesn't work.

The basic construction of this simple example has found broad application to all manner of problems in economics. I have stressed applications to industrial organization because those are the applications I know best, but in almost every one of the problem areas for credible threats or promises listed in the previous section, one can analyse credibility based on reputation constructions of the sorts discussed here.

The importance of what players know about others

Consider the game of complete and perfect information depicted in Figure 4.12. This game is called the centipede game, and it is the creation of Rosenthal (1980).

This is a game of complete and perfect information and can be solved by backward induction. At the last node, player B will choose d, since that gives him 101 instead of 100. Hence at the next-to-last node, if A chooses R she will net 98, and thus she prefers D, which gives her 99. Hence at the next-to-next-to-last node, B chooses d, getting 100 instead of the 99 he will get if he chooses r. And so on. Applying backward induction, we 'predict' that A will begin the game with D, getting 1 and giving 1 to B.

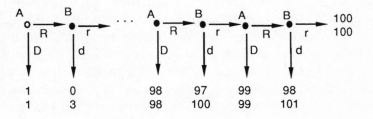

FIG. 4.12.   The centipede game

This is a pretty bad prediction empirically. I have played variations on this game with students, and it is rare indeed that A begins with D. More careful empirical evidence is supplied by McKelvey and Palfrey (1990).

This conflict between the theoretical predictions and empirical evidence is closely related to similar predictions encountered in finite-horizon versions of the 'reputation games' of the previous section. Those constructions, as we noted, depend on the infinite horizon; if at any point it is known that there is an upper bound on the number of stages or rounds, then the reputation constructions (theoretically) collapse. But evidence both careful (Selten and Stoecker 1986) and casual suggests that reputation-like behaviour can be observed in such situations. I will deal here with the centipede game as it is a bit easier to discuss, but much of what follows extends to these other, more realistic (but still very stylized) settings.

When asked why she is willing to play R at the first move in the centipede game, the typical A responds somewhat along the following lines. (Think of the payoffs as being in pence.) The loss of one unit is not too great a risk to take, and no matter what she does subsequently she comes off no worse than even if B will play r at his first opportunity. And the potential gain from playing R is substantial—up to a gain of 99. Perhaps B is simply a co-operative soul, who for a while (at least) is willing to continue playing r. Not being sure, A might wish to 'experiment' to find out. If there is even one chance in, say, ten, that B will choose r at the first opportunity, then this could be a very profitable speculation. And, continues a somewhat more sophisticated A, even if B is not in the least bit 'co-operative' but is interested only in maximizing his own payoffs, it is in his interests to act co-operatively, to give me the opportunity to co-operate with him further. If the chain were only three links long, perhaps none of this would make sense. But with two hundred or so links, why not give it a shot?

Of course, this loose intuition runs smack against the tight deductions of backward induction, and so we can (and do) wonder if there is any way to confirm this intuition as either sensible or nonsense. Thanks to a very important methodological innovation, the notion of a game of incomplete information, one can provide confirmation of this as being somewhat sensible or, rather, as not being nonsensical.

In the rationalization for choosing R that we have put into A's mouth, a key is that she isn't quite sure what B will do. We have phrased this rather more directly than is in fact typical of the subjects of such experiments; our A says that she is uncertain whether B mightn't be the soul of co-operation. We can run down the consequences of such uncertainty by using the extensive form game depicted in Figure 4.13. Nature moves first, determining whether B conforms to money-maximizing behavior (hereafter, B is a money-maximizer), or B is the soul of co-operation. Notice how this is done formally; if B is a money-maximizer, his payoffs are as in Figure 4.12, while if he a co-operative soul, then he has the very strange payoffs in the bottom half of Figure 4.13. I will not go into details here but simply assert that in the bottom half of Figure 4.13 B will always choose r; the payoffs there given make these actions strictly dominant. Notice A's information sets; A doesn't know at the outset whether B is a money-maximizer or not, and the only indications that A might subsequently receive are B's actions. Finally, note the probabilities given for Nature's actions; we suppose that there is a one in a thousand chance that B is a co-operative soul.

The game depicted in Figure 4.13 entails one (important) thing more. In the analysis of any game, it is always a maintained assumption that players in the game assume that other players conceive of the overall situation as they do, where the overall situation is the extensive (or strategic) form written down. So, in this instance, not only is A unsure whether B is a money-maximizer or not, but B is aware of this uncertainty

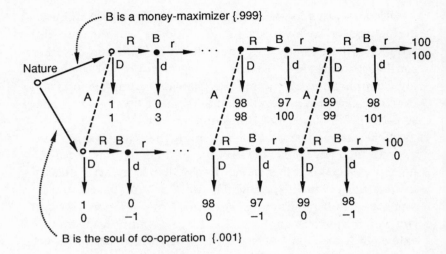

FIG. 4.13.   The centipede game with incomplete information

in the mind of A, A knows that B is aware of this uncertainty, and so on.

This is important because B may play to this uncertainty; he may act differently from the way he otherwise would because A is unsure about him. And this change in how he acts may cause A to act a bit differently, and so on. In the case of the game in Figure 4.13, precisely this happens. The analysis is very complex (even for textbooks), but this game has a Nash equilibrium in which for the first 160-odd moves, A plays R and B, whether a money-maximizer or not, responds r. (The last 40 or so moves involve a fairly complex interaction between the two parties.) Indeed, this is the unique 'credible' Nash equilibrium.[17]

---

[17] For the very sophisticated reader: That is, in this game, where A moves first, there is another Nash equilibrium in which A chooses D at the first stage because a money-maximizing B would choose d in his turn. But such behaviour by B is not sequentially rational for the reasons to be given in the next paragraph; this other equilibrium is not sequential. If you do not like the concept of sequential equilibrium, then modify the picture so that there is

While it is too hard to say why such behaviour gives the unique credible equilibrium of this game, we can perhaps communicate some of the intuition by saying why the following is not credible: player A chooses D at every opportunity; player B chooses d at every opportunity if he is a money-maximizer and chooses r if he is co-operative. To see why this is not credible, imagine that you are A and by some mishap you have mistakenly chosen R, and then B responds with r. Will you really continue with D? If you think B plays as outlined, of course you won't; you now have perfect evidence that B is a co-operative sort, and you can do much better by playing R until near the end. Now suppose you are a money-maximizing B and by some mishap A has chosen R to start the game. Will you choose d? Of course not. By the reasoning just given, if you choose r, A will think you are co-operative, and she will choose R for quite a while, which is good for you. Finally suppose you are A at the start of the game. By the argument just given, if you give the move to B, he will respond with r. You can't be hurt by choosing R, and so your (supposedly optimal) initial choice of D must be wrong.

If that argument didn't quite penetrate, let me put it this way: As soon as A is uncertain about B and B knows this, B has some incentive to take advantage of that uncertainty. In this case, B takes advantage by acting co-operatively, and so A has an incentive to give B the opportunity to 'take advantage'. Indeed, we can imagine a situation more complex than Figure 4.12, where B is a money-maximizer, A knows for sure that B is a money-maximizer, but B doesn't know for sure that A knows that B is a money-maximizer. Then 'co-operation' ensues for most of the game. If A chooses R, B thinks it might be because A is unsure that B is a money-maximizer, and so B responds with r. And even if A is sure that B is a money-maximizer, since B will respond with r, A chooses R to begin.

incomplete information about A instead of B, and then the first equilibrium described is the unique Nash equilibrium.

This may sound rather like gobbledy-gook; it certainly may sound quite far removed from anything approaching common sense. But I think there is some substantial intuition here. In the centipede game and situations like it in which players move repeatedly and for fairly small stakes in each turn, a small amount of uncertainty about what each player knows about the others can be very destructive of easy conclusions arrived at by assuming that no such uncertainty exists.

This is a success of game theory, although in a somewhat negative sense; it gives us information about the limits of the theory as a tool of practical analysis. This example, and others similar to it, indicate that in some situations the theoretical conclusions reached can be reversed by small (and quite realistic) changes in the formulation of the situation that is used. We mightn't wish to predict that players playing the centipede game will play according to the unique credible Nash equilibrium alluded to above; the prediction of that equilibrium may be as susceptible to small changes in the formulation as is the initial prediction of backward induction applied to Figure 4.12.[18] But this example and others like it have helped to resolve a paradox between common sense and empirical evidence on the one hand and overly zealous application of theory (to overly simplistic models) on the other. The empirical evidence (and common sense) made it clear that we didn't understand what was happening in games like the centipede game; the sort of analysis sketched here helps to indicate the reasons for this lack of understanding.

Interactions among parties holding private information

We have so far discussed how game-theoretic techniques permit us to model and explore dynamic competitive interactions.

---

[18] In some cases like the centipede game, however, strong predictions may still be forthcoming. The sophisticated reader can consult Fudenberg and Levine (1989) for an excellent example along these lines.

Game-theoretic techniques have also been very fruitfully applied to competitive situations where the parties involved hold private information. In general, so-called *information economics* concerns these matters, and much of the early work in information economics was undertaken without the concepts or terminology of game theory. However game theory has added a good deal to this subject. For example, in contexts of market signalling and screening, game-theoretic concepts have helped us to understand the differences between the signalling model of Spence (1974) and the screening model of Rothschild and Stiglitz (1976); see Stiglitz and Weiss (1991) for an analysis of the distinctions. Moreover, game-theoretic notions have been used to sharpen considerably the predictions of these models (Banks and Sobel 1987; Cho and Kreps 1987; Hellwig 1986). Game theory is also used in agency models, in models of search, and in studies of price formation in various institutional settings.

To illustrate how game-theoretic techniques have been applied in such contexts, I will briefly discuss one of the richest and most influential applications of non-cooperative game theory to economics, competitive auctions, and in particular the so-called winner's curse.

The story usually told about the winner's curse concerns an auction held by a government for the mineral rights to a tract of land, say, on the continental shelf. The players in the 'game' are the major oil-firms (who, in fact, form into teams for such auctions). The government might set the terms of the auction as follows: The firms must register sealed bids at a given time, specifying a lump-sum payment that the firms are willing to make for these rights. The highest bidder wins the rights and must pay the lump sum that it bid. The highest bidder must also pay to the government a certain percentage of the market value of the oil it extracts; this royalty rate is fixed in advance. Of course, the firms involved are uncertain about the existence and abundance of oil or natural gas in the tract; they base their estimates on geological data they may be able to acquire, including data they may have accumulated from experience exploring adjacent or nearly adjacent tracts. Each of the firms has its own information, and each must formulate its bid based on that information.

Of course, the more optimistic is the information of a firm, the more it is willing to bid. This then is the source of the so-called winner's curse. Suppose each firm would be better informed if it held both its own information and the information held by its peers. Because optimistic information results in higher bids, the highest bid is more likely to come from a firm with relatively optimistic information, relative to the information held by the other firms. Thus if the winning firm had all the information held by its fellow firms as well as its own, it would tend to be relatively less optimistic than it is based on its information alone. And hence if it bids solely on the basis of its own information, without taking this effect into account, it will tend to bid too much; the winner of the auction will find that winning is a curse.

This story can be illustrated and studied with the following highly stylized game. Imagine that I possess an envelope in which there is some amount of money, between $0 and $25. You, and nine other individuals against whom you will bid, do not know this amount; each of you believes a priori that each amount (to the penny) between $0 and $25 is equally likely. (The amount of money is something like the unknown economic value of the oil and gas.) I give to each of you ten a 'signal' about the amount of money in the envelope; specifically, if x is the amount of money in the envelope, then each of you receives a signal of the form $y_i = x + \epsilon_i$, where the subscript i refers to which of the ten players the signal belongs and the ϵ_i are independent normally distributed error terms, each with mean zero and variance, say, $2. Suppose that you are person 7 out of the ten and your signal is $y_7 = \$16.57$. If you undertake Bayesian inference (or simply apply common sense, if you don't know Bayesian inference), you should conclude on the basis of your information alone that the amount of money in the envelope is most likely to be $16.57. If you do know Bayesian inference, you should find it straightforward to conclude that, based on your information, the posterior distribution for the amount of money in the envelope is (to a very good approximation) normal with mean $16.57 and variance $2.[19]   In terms of our story of mineral-rights auctions, you (of course) are one of the oil companies, and this posterior distribution is your estimate of the value of the oil and gas under the ground, based on your private information.

All ten of you are invited to submit independent sealed bids for the contents of the envelope; the contents are awarded to the high bidder for the amount that person bids. Based on your information, you think the amount of the money in the envelope has an expected value of $16.57. You don't want to bid that much since then you will not make a profit if you win the auction. Perhaps you should bid, say, $1 less than your estimate, or $15.57, so if you win the auction you expect to make $1.

The statement just made may seem logical, but it is not correct as long as your fellow bidders are also competing for the envelope. This is easiest to see if we imagine that every one of the ten individuals bids $1 less than his or her best estimate of the amount of money in the envelope. In this case the person who wins the auction will be precisely the person with the highest, most optimistic piece of information. This person will be bidding $1 less than his or her information, which on average will be considerably more than the average of all the pieces of information, which (in this very simple setting) is the best guess one can make as to the amount of

---

[19] Because the prior is uniform, it is approximately diffuse. It is only when you get a signal close to one of the two end-points of $0 and $25 that things become more tricky; you must deal with truncated normals.

money in the envelope. In this setting, if everyone bids $1 less than his or her signal then the winner will (on average) be quite cursed.[20]

So how should you bid in this situation? We can think of this as a game in which nature moves first, generating (randomly) the amount of money in the envelope and also the ten signals. Then the ten players simultaneously and independently formulate their bids, each knowing his or her own signal. (Each bidder has as many information sets as there are possible signals for that bidder.) The game is too complex to depict, but with somewhat different parameterizations it is possible to 'solve' it analytically.[21] A strategy for each player is a bidding function—how much do you bid if your signal is $15.57, if your signal is $15.58, $15.59, and so on—and in some cases it is possible to produce a Nash equilibrium in these bidding functions.[22] Moreover, one can go on to consider how changing the rules of the auction might affect its outcome (and government revenues), what is the effect of having some

[20] The winner will, on average, lose approximately $1.20; the computation of this figure is not a trivial matter.

[21] See, e.g., Wilson (1977).

[22] While we cannot solve analytically for Nash equilibrium strategies in this particular parameterization, we can compute the equilibrium to any desired degree of approximation. The details are fairly complex, but you can get a rough idea of the equilibrium with the following computations. As long as your signal is not too extreme (close to or less than zero, or close to or more than $25), your strategy is approximately to bid a fixed amount k less than your signal y_i. All your opponents will use the same bid function, so you win precisely when you have the largest signal; if you condition on y_i being the largest signal out of ten, then you can compute the expected amount of money in the envelope as $y_i - K$ for some other constant K. Hence your expected profit is $(0.1)(k - K)$, or the probability (one-tenth) that you win the auction times your expected winnings if you win. If you decrease your bid by δ, you decrease the probability that you win by an amount that depends on the joint distribution of the first- and second-order statistics of a sample of ten independent normals (the chance that the second-order statistic is within δ of the first), and you increase the expected amount you win. The equilibrium is (roughly) that value of k for which this variation in your strategy gives (to a first-order approximation) no change in your expected value. Of course, near the edges things are much more complex.

firms with better information than others, and so on. And (as you might have guessed) one can study auctions with a dynamic element to them—sequences of auctions, where winners of early auctions (say) become better informed about the value of objects that will be auctioned later.

Concluding remarks

The first and perhaps most important role that game theory plays in economic analysis comes in the formulation and framing of issues. The language of extensive form games permits economists to ask questions about the dynamics of competitive interactions; the language of other parts of non-cooperative game theory—so-called games of incomplete and/or imperfect information,[23] examples of which are the games in Figures 4.9 and 4.13 and the auction game just described—are well suited to pose questions about competitive interactions in which parties have proprietary information. Indeed, these modelling techniques have focused attention on issues of dynamics and proprietary information, framing the issues so that the 'mechanics' of the interaction—who does what when with what information—come to the fore. If one believes that the mechanics or institutional forms of interaction matter, as do I along with many others, then simply focusing attention and framing debate in this way is a notable success.

As for the specific solution techniques of and the insights contributed by game theory, I contend that the major successes have come primarily from formalizing common-sense intuition in ways that allow analysts to see how such intuitions can be applied in fresh contexts and permit analysts to explore intuition in and extend it to slightly more complex formulations of situations. The four specific examples recounted in this chapter

[23] A formal distinction is sometimes made between games of *incomplete* information and games of *imperfect* information, but this distinction is both a bit arcane and of little significance for applications. Thus I use the terms interchangeably here.

are, in my opinion, among the most important applications of the techniques of non-cooperative game theory to economics. But they were chosen for inclusion here as well because each makes the point that game theory has succeeded when it begins from a common-sense observation and takes a few small steps further along.

In a story about famous young economists, *The Economist* newspaper (24 December 1988) wrote about one of the tribe of economists who employ game theory, Jean Tirole, as follows: 'Drawing on game theory and other strange techniques, [Tirole's] approach began to make sense of strategic behaviour that had seemed theoretically unmanageable.' The choice of modifiers is good in places and perhaps misleading in others. It was the 'theoretical' unmanageability of otherwise reasonably intuitive behaviour that was the issue; non-cooperative game theory provided the few relatively simple tools needed to frame and analyse these behaviours in a mathematical theory. But are the techniques 'strange'? If strangeness is taken here to connote novelty, then the word is apt. But the terminology and techniques employed are not or rather should not be at all alien to one's common-sense thinking about a situation.

Few if any of the conclusions of successful game-theoretic analyses are startling or mysterious or arcane; after the fact, it is usually easy to say, 'Well, I think I already knew that.' Of course, what one remembers having known before an explanation was offered is often more than what one really did know. Perhaps it is more truthful after the fact to be saying 'I knew that subconsciously' or 'I should have known that; it's so obvious.' But that is the level at which successful game-theoretic analysis works. It contributes (*a*) a unified language for comparing and contrasting common-sense intuitions in different contexts (if X is sensible in context A, then it is as well in contexts B and C); (*b*) the ability to push intuitions into slightly more complex contexts (if X is common sense in context A, then perhaps X' in context A' is as well); and (*c*) the means of checking on the logical consistency of specific insights and, illustrated especially well by the discussion of the centipede game, a way of thinking through logically which of our conclusions may change drastically with small changes in the assumptions (if X is predicted in context A but doesn't seem such common sense, then we might learn a great deal by seeing that not X is consistent behaviour in context A').

These are not attributes unique to game-theoretic techniques. Much of the benefit of mathematical modelling and analysis in economics generally stems from these sorts of contributions. But if this is true in general about many of the useful mathematical models in economics, it is, I contend, especially true for virtually all of the game-theoretic models that have successfully contributed to our understanding of economic phenomena.

Some bibliographic notes

I would be remiss not to give names of some of the individuals who pioneered the game-theoretic techniques explored here. The backward induction procedure goes back to the very origins of the subject (Zermelo 1913), but in terms of more modern developments, and especially the extension of backward induction-like ideas to more complex games, Reinhart Selten is the foremost pioneer. In his seminal work on 'perfecting' the notion of Nash equilibrium (Selten 1965, 1975), he provided the basic machinery for examining the credibility of threats and promises. He has also been a leading force in getting economists to think through problems dynamically.

As noted earlier, the folk theorem is widely attributed to the 'folk literature' of game theory. But James Friedman (1971, 1977) was very early in applying these ideas to implicit collusion among oligopolists, and Robert Aumann and Lloyd Shapley (1976) provide the earliest written formal statement of the theorem of which I am aware.

In the last elaboration on the story of the entry-deterring

monopolist we modelled uncertainty in the mind of the entrant about the costs of the monopolist; in the variation on the centipede game there was uncertainty in the mind of one player about the 'character' of the other. These models are examples of *games of incomplete information*. Due to John Harsanyi (1967-8), this methodological innovation—a way to model situations in which players are uncertain about the characteristics or capabilities of their rivals—has been of enormous importance in the recent literature of economics by extending the range of situations that can model led and analysed; it is central to many applications of game theory to the study of information economics.

If you wish to read further about game theory as it applies to economic analysis, a number of textbooks have begun to appear (and more are on their way as I write this). Kreps (1990) and Rasmussen (1989) provide detailed accounts of the basic notions, and both of these books provide links to the literature of information economics. Tirole (1988) gives a shortened 'Game Theory User's Manual' as an appendix, and his book is particularly valuable as a source for all manner of applications of these techniques to problems in industrial organization. In particular, he gives a very detailed discussion of entry deterrence and implicit collusion in oligopoly. For readers specifically interested in auctions and competitive bidding from a game-theoretic perspective, the surveys of McAfee and McMillan (1987) and Milgrom (1989) are recommended.

5
The problems of game theory

In this chapter, I will explore a few of the deficiencies of game theory as a tool for modelling economic phenomena. The four deficiencies that I will point to have been selected because they are related to the strengths that were the subject of the preceding chapter, and also because for the most part they are weaknesses that share a common root and so can be attacked together, or so I believe and will attempt to convince you in Chapter 6. In any case, this list of problems is not exhaustive.

Some readers of this chapter have complained that the tone is too negative and that I am too tough on game theory. In particular, of the four deficiencies that I cite, three are mainly sins of omission and only one is a sin of commission. That is, in three of the four cases we confront issues that game theory has not addressed. One can argue that these issues cannot naturally be addressed by game theory; asking that the theory do so shows a misunderstanding of what the theory is good for. That may be, but I hope you will be convinced that from the point of view of economic applications there are important gaps in our arsenal that ought to be addressed, whether by a subject called game theory or by tools that are given some other name. Notwithstanding this, let me assure you that I believe what I wrote in Chapter 4; viz. game theory has helped us to make important contributions to our understanding of economic phenomena. Any criticisms that I go on to make are made from that starting-point.

One other word of introduction is appropriate. To illustrate some of the deficiencies of game theory, I will use aspects of the theory of bargaining. Do not conclude from this that I am denigrating the work that has been done on this particular subject. Bargaining is an extremely difficult topic because in many settings it runs right up against the things game theory is not so good at. It is a measure of the extraordinary recent progress that we are able to see much better today than ten years ago in what directions to proceed.

The need for precise protocols

Imagine two individuals, players A and B, who are set the task of bargaining over how to split between themselves some economic good. To be very precise, the two are put in a room with Figure 5.1(a) and a blank 'contract' of the sort depicted in Figure 5.1(b). They are told that they have thirty minutes to negotiate. If they come to agreement, they can have any of the pairs of points in the shaded area of Figure 5.1(a) as payoffs (in units, say, of dollars); i.e. because the point (5,4) is in the shaded region, they can agree to $5 for player A and $4 for player B. But since (5,6) is outside the shaded region, they cannot have $5 for A and $6 for B. They will be left in the room for thirty minutes. At the end of that time a referee will come into the room and collect the contract. If the contract has been filled out with a 'proposal' that is feasible according to Figure 5.1(a) and if the contract is signed by both parties, then the proposal will be implemented. If the contract is filled out with an unfeasible proposal or if either or both parties have failed to sign the contract, the two get one dollar apiece. (Forgery is not an issue.)

This is a highly stylized version of an important problem in economics. In special cases, economic theory makes fairly bold predictions what will result from such bargaining. Almost everyone would agree, for example, that neither player will accept an outcome worse than what he or she can get if there is no agreement. So if, in the example, disagreement results in payments of $4.90 to each player, then there isn't much left to

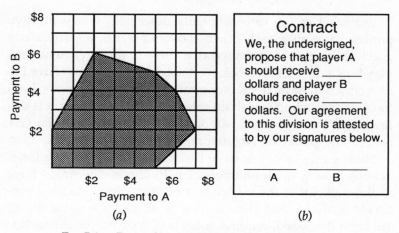

FIG. 5.1. Props for a simple bargaining situation

bargain over. Roughly put, the theory of perfect competition is built around such a story; what an individual gets if he doesn't come to agreement with another is represented by what he can get by taking his business elsewhere; when markets are competitive, there are many 'as good' or almost as good alternative trading partners for each individual; and market prices (known to all) establish the most anyone can get at those alternatives. Or suppose that we imagine a single player A who is put in a room with many players B, and the rules are that A can conclude an agreement with at most one player B, each of whom can conclude an agreement with A and no one else. Then we would expect (for the case of Figure 5.1(a)) that A could use her 'monopoly' position to obtain approximately $7 for herself and $2 for some lucky B. Although the situation is more complex, the notion that a seller can set her price when she alone sells the good and there are many (more or less equivalent) buyers traces from such arguments.[1]

---

[1] We should go slow here. The models of competitive or monopolistic markets may give us good predictions in these cases, but empirical or experimental verification should be sought. And we might wonder whether fine-grained

But what of the case originally described, where the two parties have only each other and neither has a particularly good alternative to pursue (i.e. if there is no agreement each player gets $1)? In textbook economics, this is often referred to as a situation of bilateral monopoly, and most textbooks don't have very much to say about this problem except that it is difficult and that outcomes will depend only partly on economic factors and partly on negotiating skills and perceptions.

Since game theory is meant to apply to competitive interactions among small numbers of individuals, one might hope at the outset that the theory will fill this gap in classical economic analysis. But game theory, or at least the game theory we have discussed, requires models of competitive interaction that are very exact and precise as to the strategies available to the players. Our story about sticking A and B in a room and telling them that the referee will appear in thirty minutes doesn't establish very much of an extensive form game, or rather it establishes a game so incredibly rich and full of possible moves and countermoves that we could not hope to analyse it in the fashion we have been using.

This then is the first problem on our list. *Game-theoretic techniques require clear and distinct 'rules of the game'.* Analysis of free-form competition such as this bargaining situation is

models of the institutions of bargaining in such contexts can be analysed in a way that will give insight into why these are good predictions. For example, the intuition for the monopoly outcome in the many B, one A situation probably depends in some measure on the fact that the one A can deal with only one of the many Bs. What would you predict if there were one A and many Bs and A can come to agreement with as many Bs as she wishes? What then is the case for the classic monopoly outcome when the monopolist has a constant average-cost production technology? Game-theoretic techniques have been applied in the literature to precisely these questions: What are the bargaining-theoretic foundations of competitive equilibrium and monopoly power? Which specific bargaining institutions would support (according to theory) such outcomes, and which would not? For a less technical introduction to some of the work on these questions, see Rubinstein (1989). More technical treatments include Ausubel and Denekere (1989), Gale (1986), Gul, Sonnenschein, and Wilson (1986), and Osborne and Rubinstein (1990).

not within the realm of the techniques provided. In this case we must be more precise about the protocol of bargaining.[2]

Let me repeat from the introduction to this chapter that it is not unreasonable to rebut this criticism with the assertion that the point of non-cooperative game theory is to see how outcomes change with changes in precise protocols. Put a bit too flamboyantly, game theory does not help us (so far as I know) in resolving the mysteries of quantum mechanisms, in devising a cure for cancer, or in composing a sonata, but it would be harsh indeed to say that this constitutes a deficiency of the theory. It isn't intended to do such things. Similarly, one can take the position that non-cooperative game theory is a tool for studying the effect of protocol on outcomes; dealing with situations in which the protocols are imprecise is outside the ken of the discipline. I respond to this rebuttal by saying that if game theory is necessarily so limited, then it is extremely limited as a tool of analysis for *some* important economic problems. And I would prefer that we not give up hope, but instead think in terms of how we might adapt the theory so it is better able to cope with ambiguous protocols.

Too many equilibria and no way to choose

So that we can proceed with a game-theoretic analysis of bilateral bargaining, let us be precise about the bargaining protocol. There are many possibilities, among which is the following simple demand game: The two players simultaneously and independently demand the amount of money they wish. These

---

[2] We might still hope for results that say that the results of analysis are relatively insensitive to the bargaining protocol, at least if the bargaining protocol is not too one-sided; from this we might gain the confidence to claim that although we can't write down the precise protocol for our 'lock them in a room for thirty minutes' situation, we expect in this unstructured situation an outcome similar to those obtained in our wide range of precise protocols. As we shall see, the extant bargaining literature does not seem to lead us to such a happy conclusion.

two demands are compared, and if they are compatible (lie in the shaded area), they are implemented. If not, the game is over; there is disagreement.

Or, to complicate this game a bit, if the two initial demands are incompatible, then the two players (having heard those demands) simultaneously and independently make a second round of demands; if these are compatible, they are implemented; and if not, a third round of demands are made, and so on, for (say) up to ten rounds of demands. If after ten rounds there has been no agreement, then the disagreement outcome of $1 apiece is implemented.

Or, to complicate matters further, suppose the two players are involved in some complex (precise) scheme of bargaining; say A makes a proposal first (specifying what each will get) to which B can agree or not; if not, then B has the opportunity to make two proposals either of which A can accept; then they make simultaneous demands; then . . . Put in anything you wish, as long as it is finite and ends with the following: And, after all that, if there is still no agreement, then A and B make one final simultaneous pair of demands. If these final demands are compatible, they are implemented. If not, the disagreement outcome is implemented.[3]

The punch-line to this is that in all of these situations, any *feasible, efficient, and individually rational division* is the outcome of a Nash equilibrium of the bargaining game. Let me define terms. A *division* is an amount of money for A and an amount of money for B. A division is *feasible* if it lies within the shaded

---

[3] Indeed, the argument we will give applies imprecisely to the following somewhat vague game: Put the two in a room for thirty minutes with the contract. If they have filled it out, implement their agreement. Otherwise give them one more chance with simultaneously rendered demands. Our statement about the need for precise protocols should be modified to read: As long as protocols reserve certain rights to the player (in this case the right not to agree until the end at no cost), and if the 'game' ends in a particular fashion (in this case with simultaneous demands if no agreement was reached previously), then we may be able to proceed with some analysis.

region of Figure 5.1(*a*), *efficient* if there is no feasible division that gives more to one player and as much to the other, and *individually rational* if it gives each player at least as much as the player can guarantee for himself, in this case $1.

The proof of this statement is quite simple. Take any feasible, efficient, and individually rational division, say $6.50 for A and $3 for B. Imagine that A plays the strategy: Ask for $6.50 at every opportunity, and (in variations where this is possible) turn down any proposal that gives anything less. Player B's best response to this is to ask for $3; he isn't going to get any more, and there is no point in taking any less. (In particular, there is no point in taking $1 by causing a disagreement.) And similarly if B asks for $3 every time and refuses to take any less, then A's best response is to ask for $6.50.

There are, therefore, lots of Nash equilibria to this game. Which one is the 'solution'? I have no idea and, more to the point, game theory isn't any help. Problem no. 2 of game-theoretic techniques is that *some (important) sorts of games have many equilibria, and the theory is of no help in sorting out whether any one is the 'solution' and, if one is, which one is.*

Simultaneous-offer bargaining is only one context where there is this problem of too many equilibria and too little guidance for choosing among them. Another very important context is that of repeated interaction and reputation. To take a concrete example, imagine two individuals who each day play each other in the prisoners' dilemma game, depicted here in Figure 5.2(*a*). Imagine that each plays in order to maximize his or her discounted sum of payoffs, discounting with a discount factor 0.9. (That is, if $\{u_1, u_2, \ldots\}$ is the infinite stream of payoffs obtained by one of the two players, where u_n is the player's payoff in round n of play, then this player seeks to maximize $\sum_{n=1}^{\infty} (0.9)^{n-1} u_n$.)

Although we didn't discuss precisely this repeated game model in Chapter 4, it shouldn't be hard for you to see, based on arguments given before, that we can sustain 'co-operative

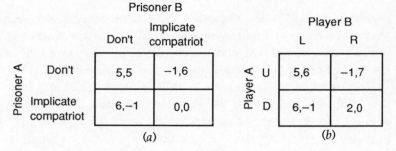

Fig. 5.2.   The prisoners' dilemma game and a variant

behaviour' as a non-cooperative equilibrium in this situation.[4]
Each prisoner adapts the following strategy:

> Don't implicate the other, as long as neither has impli-
> cated the other previously. If either implicates the other,
> then implicate the other in every subsequent round.

This is an equilibrium because each prisoner would rather
keep co-operation alive for a present value of future payoffs of
50 than take advantage for one period and then suffer there-
after for a present value of future payoffs of 6. In essence,
the promise to co-operate is credible because of the threat to
punish non-cooperative behaviour. And the threat to punish non-cooperative behaviour is credible because once non-
cooperative behaviour is unleashed, there is no advantage to
trying to restore co-operation unilaterally; the other party will
not respond (according to the strategies given).

This is but one Nash equilibrium in this context, however.
It is also equilibrium behaviour for each player to act non-
cooperatively at all times. More interestingly, it is Nash equilibrium behaviour for each player to alternate co-operation and
non-cooperation, giving each a present value sum of payoffs of
approximately 26.32. The following two strategies constitute
yet another Nash equilibrium:

---

[4] 'Co-operative behaviour' here means co-operation between the two prisoners and *not* with the authorities.

Player A alternates co-operative and non-cooperative actions, as long as player B plays co-operatively at all times (and player A herself doesn't deviate from this pattern); player A plays non-cooperatively forever in the event of any deviation.

Player B plays co-operatively at all times, as long as player A alternates between co-operation and non-cooperation (and player B himself doesn't deviate from co-operation); player B plays non-cooperatively forever following any deviation.

If player A begins with co-operation, this pair of strategies yields net present values of (approximately) 54.74 for A and 21.58 for B. Why does B put up with this? Because even though A does better than B, B does better than if he deviates. By deviating he will at best get 6 immediately and zero thereafter.

(Why should player B ever agree to such an arrangement? Suppose that the payoffs in the game are as in Figure 5.2(*b*). Now we can imagine player A objecting to the equilibrium in which both sides co-operate in each round as follows: She (A) gets only 5 while he gets 6, and if co-operation breaks down, then she will get 2 while he gets 0. Since she has a bigger threat, perhaps she deserves to do better. The point is that if the game is symmetric, as in Figure 5.2(*a*), both we and the players involved might find asymmetric equilibria somewhat unintuitive. But if the game is asymmetric, then asymmetric equilibria do not seem so unreasonable.)

And so on. Essentially, as long as a pair of payoffs is feasible and leaves each player with enough at stake so that neither wishes to deviate and set off punishment, then that pair of payoffs can be sustained as a Nash equilibrium in this context. And while we have illustrated this point in the context of repeated play between two long-lived opponents, similar (but slightly more complex) things happen with one long-lived opponent facing a sequence of short-lived opponents, as in the tale of the employment relationship told in Chapter 4.

On what basis is an equilibrium chosen when one is chosen?

Why do we worry about games that have many equilibria? It is *not* multiplicity *per se* that is a problem. The cities game of chapter 3, for example, has many Nash equilibria—128 at least[5]—but I am reasonably confident that players will select one equilibrium in particular.

To take a second example, consider the following particular implementation of the battle of the sexes game depicted in Figure 5.3(*a*), which is reproduced from Figure 4.2 although the labels about ballet and boxing have been removed. The two players will be chosen at random from a given population, say from a group of university students. One player will be selected first and asked to choose between the first and second columns, recording his or her choice on a piece of paper that is given to the referee. Then, after the first player's choice is recorded, the second player is selected at random from the group and is asked to choose between row 1 and row 2. The choice of row by the second player and the recorded choice of column by the first are compared and payoffs are made. This entire procedure is explained carefully to all the students before any actions are taken.

From the perspective of game theory, the fact that player B moves first chronologically is not supposed to matter. It has no effect on the strategies available to players nor to their pay-offs. It does give the two a piece of information on which to co-ordinate their choices, however, and my own casual experiences playing this game with students at Stanford University suggest that in a surprising proportion of the time (over 70 per cent), players seem to understand that the player who

---

[5] There are 128 pure-strategy equilibria, corresponding to the 128 possible ways to partition the seven unassigned cities into two sets. There may be mixed-strategy equilibria as well; if we consider the variation in which a player gets one dollar per point if he alone lists the city and loses two per point if both do, and if we drop the bit about doubling winnings if the two lists precisely partition the nine cities, then there are uncountably many mixed-strategy equilibria.

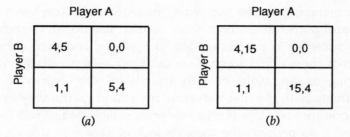

Fɪɢ. 5.3. Variations on the battle of the sexes game

'moves' first obtains his or her preferred equilibrium. Substantial caveats are in order here; of all the casual experiments I report in this book, this is the most casual (or, rather, this is the one where I believe there is the greatest danger that discussion in class preceding play of the game has contaminated the results). Still, the result has some intuitive appeal. And, if I may be allowed a conjecture, I suppose that this effect will be stronger in the case of the game in Figure 5.3(*a*) than it will be in the game in Figure 5.3(*b*) because in 5.3(*b*), the 'inequality' in the two equilibria is much larger.

The point is that in some games with multiple equilibria, players still 'know' what to do. This knowledge comes from both directly relevant past experience and a sense of how individuals act generally. And *formal mathematical game theory has said little or nothing about where these expectations come from, how and why they persist, or when and why we might expect them to arise.* The best discussion of these sorts of things (at least in the literature of game theory) remains the original treatment due to Thomas Schelling (1960); little to no progress has been made in exploring Schelling's insights. Indeed, as we see in the case of the battle of the sexes, things that game theory tells us to ignore actually can be the very keys upon which players co-ordinate.

No, that is too harsh. Imagine playing the battle of the sexes in which a pair of players is selected at random from a large population; then the two move 'in sequence' (except that the second player doesn't see what the first has done); the results are revealed to the population as a whole; and then another pair is chosen, and so on. One can imagine a population that, owing to the results of early rounds of play, 'learns' to co-ordinate with the first mover in any round giving way to the second mover. Once such a custom is established, it will likely persist; so perhaps the theory's lack of choice in this case is warranted; either 'custom' is a possible outcome. Or, to make a completely untested conjecture, perhaps the results of my casual experiment would be reversed if the subject population was a group of students from another culture in which defer-ence to peers was conventional. Or imagine a subject popula-tion in which one participant was always, say, a Korean student and the second always a Korean professor; these identities be-ing known to the participants. My limited experiences with Korean students leads me to conjecture rather confidently that the temporal order of moves will be almost completely disreg-arded and another rule will be used in its place. Then if the theory doesn't tell us *how* to select among the equilibria, it is good that it identifies all as possible solutions.

Still, without (until next chapter) prejudging the possibility of a theory of equilibrium selection based on the types of con-siderations outlined above, it is safe to say that formal game theory has not delivered such a theory as of yet.

Is equilibrium analysis inappropriate, and, if so, what then?

Now return to the examples of games in which there are many equilibria and neither theory nor common sense gives us (as outside observers) a sense of how to choose among them. Sim-ultaneous-offer bargaining over the region depicted in Figure 5.1(*a*) is one example; repeated play of the game in Figure 5.2(*b*) is another. Might this inability to select an equilibrium infect the players as well? And if it does, how will they behave? That is, grant that we as observers cannot make a selection. Must the players involved still have enough of a common sense of the situation so they will play according to *some* equilibrium albeit one that we cannot identify *ex ante*, or might they be so 'confused' that they play according to no equilibrium at all?

At this point I would like to say that the answer to this question is clear and unambiguous. Players might very well take actions that conform to no equilibrium whatsoever. But this is a tricky question. The range of equilibria in the repeated play of Figure 5.2(*b*) is very large indeed, and it will be hard to say that any observed behaviour is inconsistent with every equilibria. If we restrict attention to equilibria in which the players choose their strategies with certainty (so-called pure-strategy equilibria), then we can get some clearly falsifiable predictions for simultaneous-offer bargaining games (that is, there cannot be disagreement unless each player makes a demand so large that the other does as well to take the disagreement point). But perhaps we should allow for randomized equilibria.[6] And

---

[6] This is the first mention (in the text) of randomized strategies, and an explanation is perhaps in order for readers who are novices to game theory. Up to now, we have considered only so-called *pure strategies*; a player decides on a single course of action and carries it out. But in some contexts players may choose their course of action somewhat randomly. The canonical example of this is bluffing in poker. If you hold a bad hand, you will sometimes bet heavily on it and sometimes not, choosing (in each instance) randomly between bluffing (betting) and not. The idea is that you don't want your betting behaviour to signal to your opponents what cards you hold; you randomize between bluffing and not so that when you bet heavily, your opponent is confused as to whether you hold a good hand or not. Moreover, in an equilibrium you would choose between bluffing or not with a bad hand just enough so that your opponent is indifferent between calling your bluff or giving in to it; if your opponent always calls your bluff, then you would do better never to bluff, while if your opponent always folds when you bid heavily, you would do better to bluff with higher frequency, to win more hands with poor cards. There are other situations in which we see individuals choosing their strategies 'randomly'; e.g. tax authorities will choose somewhat randomly who to audit, accounting firms conducting an outside audit of a client will choose somewhat randomly what to audit, and so on. Within game theory, randomized (or, as they are sometimes called, mixed) strategies play an important technical role as well; in many games they guarantee the existence of at least one Nash equilibrium. Whether any randomized strategy is really 'intuitive' (i.e. part of the

in any game, perhaps we should take into the account that players may see the game as one where they are unsure of the motives of their fellow players, so we must consider whether the observed actions conform to equilibria in games of incomplete information built out of the game structure we do see. Readers who are relative novices to the literature of game theory may not quite be following what I am saying here, so let me put it this way. Game theorists are very clever individuals, and given almost any form of behaviour, they can build models that 'explain' the behaviour as the result of an equilibrium in a sufficiently complex elaboration of the game originally written down.[7] So I must put the question this way: Can we, as outside observers, specify a priori a 'reasonable' model of the world of the players for which (we are confident) they will play some equilibrium? It is this question that is important to us in considering game theory as a tool of economic analysis; it is cold comfort (and useless theorizing) to know that there is always some explanation of behaviour consistent with equilibrium theory, but we couldn't reasonably say what the explanation is until we see the behaviour.

And having put the question this way, then I think it is intuitive to say that players in a situation may exhibit behaviour

---

evident way to play a given game) is another matter, and game theorists often defend this theoretical artifice by showing how randomized strategies can be thought of as pure strategies in games where players hold private information. (Such defences go by the rubric of 'purification theorems'; the seminal reference is Harsanyi (1973).) In the context of simultaneous-offer bargaining, in any pure-strategy equilibrium the players will result in agreement and leave no money on the table (unless each asks for so much that the other does as well by taking the disagreement outcome as by agreeing); so if we see money left on the table or disagreement, we falsify the proposition that the two are playing a pure-strategy equilibrium. But equilibria with randomized strategies can be constructed in which there is disagreement and money is left on the table, each with positive probability. So if we see either of those outcomes, how do we know that the players aren't playing some equilibrium in randomized strategies?

[7] There are even theorems about the range of things that can be explained. See, e.g., Ledyard (1986).

that conforms to no equilibrium for a 'reasonably specified' model of the situation. We may believe that each player has his own conception of how his opponents will act, and we may believe that each plays optimally with respect to this conception, but it is much more dubious to expect that in all cases those various conceptions and responses will be 'aligned' or nearly aligned in the sense of an equilibrium, each player anticipating that others will do what those others indeed plan to do.

Once we admit to the existence of this sort of disequilibrium situation, we can wonder what game theory tells us about it. What predictions do we make about behaviour in such cases? How does the theory tell us players will behave?

There are 'nonequilibrium' or 'disequilibrium' notions in formal game theory; notions such as *rationalizable strategies* fit here. But such notions are sometimes very weak, precluding very little. We might want a theory of dynamic behaviour that is weaker than equilibrium yet with greater strength than notions such as rationalizability. *Game theory does not offer much at all of this middle ground.*

Yet in many economic contexts, such a middle-ground theory is tremendously important. I have in mind cases of industrial competition when a new product arrives or an entirely new situation is created by a change in the legal or technological structure. A good example to think of is competition in the US domestic air-transport business just following deregulation of the industry. It is fairly clear that the moves and countermoves of the participants in the early days of deregulation were to some extent confused and based on quite different notions of how competition should be structured.[8] And it seems fairly clear that the eventual structure of industries can be powerfully affected by early events in which firms lock up channels of distribution, erect entry barriers, begin to form strategic expectations, and so on. Yet game theory, which has

[8] But, harking back four paragraphs, try to think of a tight empirical test of the assertion just made.

depended so heavily on equilibrium analysis, is of little help in such situations.

An aside: Co-operative game theory and bargaining

In the face of problems with too many equilibria, we might try to select among the equilibria using formal principles that compare across situations. In a sense, this turns the usual comparative statics exercise of economics on its head. Rather than see how the predicted equilibrium changes with changes in the model's parameters, we construct the equilibrium selection from our notions of what are appropriate comparative statics. This takes us into one of the realms of co-operative game theory, and while it is a bit of an aside from our main development, since we introduced bilateral bargaining it may be of interest to say a bit about how this sort of development can proceed in that context.

Think of a general class of bargaining problems that are described by variations on Figure 5.1; consider simultaneously all situations described by a (convex) set of possible agreements and, interior to the set of possible agreements, a 'disagreement outcome' that will be implemented if the players do not reach agreement.[9] Suppose we decide that in symmetric situations, we expect symmetric and efficient solutions. That is, in the symmetic situation where the problem is to split $100 and both players get nothing if no agreement is reached, we would expect a split into $50 shares for each.[10] Now add a

---

[9] This particular context has enormous historical significance to the development of game theory. The problem of too many equilibria in the simultaneous-demand game was first observed by Nash (1950), who resolved the problem by an ingenious (but somewhat *ad hoc*) procedure. Not happy with his own resolution, or perhaps looking for some other justification for the solution he proposed, Nash (1953) went on the development I am about to sketch. The relation between Nash's bargaining solution (which follows) and non-cooperative approaches to bargaining continues to be of interest to researchers and the subject of continuing work.

[10] Here and throughout my discussion of bilateral bargaining, I assume the parties are risk-neutral.

second 'rule' or principle: If in a given bargaining situation some possible agreements are removed, and if those possible agreements are not in fact agreements that we think would be reached, then their removal does not change our prediction as to what will happen. Because these possible agreements are (in our theory) not going to be agreement points, they are irrelevant. For example, suppose we modify the split $100 situation by telling the two that no agreement that gives player B more than $55 will be permitted. Since all the points that are thereby precluded are not solutions of the simple split $100 game (the equal-share split which we suppose solves the simple game is not precluded), our second principle implies that equal-share splits will still be the solution, even with this restriction. (You can decide for yourself what you think of this particular principle.) Add to this a third, somewhat more technical principle: Suppose we have two bargaining problems where the first is obtained as a positive affine translation of the other, i.e. there are constants $a > 0$ and b such that if (x,y) is feasible in the first situation, then $(ax + b,y)$ is feasible in the second, and if (x^d,y^d) is the disagreement outcome in the first, then $(ax^d + b,y^d)$ is the disagreement outcome in the second. Then (the principle states) if (x^*,y^*) is the bargaining solution of the first problem, $(ax^* + b,y^*)$ is the bargaining solution of the second. This third principle is likely to seem somewhat inane to many readers. But if we interpret the outcomes not as dollar amounts but as levels of von Neumann–Morgenstern utilities, then this principle says something along the lines of, The scale and origin of the cardinal utility function that is used to represent players' preferences should not matter to the bargaining problem. These three principles can be satisfied (for all bargaining situations) by only one 'rule', the so-called Nash bargaining solution. And if you don't like these three principles, co-operative game theory has others to offer. If you are interested, Roth (1979) provides an excellent introduction to this general approach to the bargaining problem.

Choosing among equilibria with refinements

The third problem of game theory we explore concerns so-called *equilibrium refinements*. In games with multiple Nash equilibria, one of the ways to select among them involves the invocation of some stronger notion of equilibrium.[11] Equilibrium refinements are strengthenings of the requirement that behaviour constitutes a Nash equilibrium, usually strengthenings that invoke in some manner or other the idea that players should not be allowed to make incredible threats or promises or to draw incredible inferences from things they observe. These strengthenings have been very popular in the application of game theory to economic contexts in the recent past, and in this section I wish to make the case that greater scepticism is called for in their application than is normally applied.

We saw such refinements in action in Chapter 4, but let us begin with a very quick course in how they work. In the game in Figure 5.4(*a*), both U-l and D-r are Nash equilibria. But U-l is based on the incredible threat by player B to choose l given the opportunity. As we noted in Chapter 4, we really don't expect B to play l once it is clear (as it must be if B is choosing) that A selected R, and so while U-l is a Nash equilibrium, it is dismissed as a bad prediction. The refinement being employed here is simple backward induction. We saw how this refinement selected a single Nash equilibrium out of many in the von Stackelberg game of Figure 4.3, we mentioned that an extension of this idea—subgame perfection—selected a single Nash equilibrium out of many in the game of Figure 4.7, and although we didn't give any of the details, it is a further extension of this idea—sequential equilibrium—that is applied to the game of figure 4.9.

---

[11] This is not a technique that is of much use in either the context of simultaneous-offer bargaining or the context of repeated play. But it can be incredibly powerful in the context of alternating-offer bargaining, an application discussed at the end of this section.

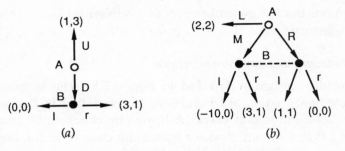

Fig. 5.4. Two extensive form games illustrating
backward and forward induction

To take a second example, in Figure 5.4(*b*), both L-l and M-r are Nash equilibria. But the choice of l by player B in the L-l equilibrium is usually held to be incredible. If player B is given the move, then his response is dictated by whether he thinks player A chose M or R. If the odds are greater that player A chose M, then r is B's best response. If the odds are greater that player A chose R, then l is B's best response. So the question is, Is it credible for B to suppose that A is more likely to have chosen R, given that L was not chosen by A? Since the best that A can obtain by choosing RA—a payoff of 1—is less than the payoff of 2 that A can obtain for sure by choosing L, and since M offers the chance for A of obtaining a payoff of 3, it is often asserted that B cannot credibly hold beliefs that justify the choice of l; B will respond to 'not L' with r, which in turn leaves us with the M-r equilibrium only. In the literature, this sort of argument is loosely referred to as *forward induction*; when extended it can be very powerful in the context of signalling games, for example.

These are but two examples; the recent literature is full of papers both touting and using refinements to select among Nash equilibria. Indeed, as we just noted, in our earlier discussion of variations on the story of the entry-deterring monopolist we used the logic of these refinements to settle on one of the several Nash equilibria as our prediction of what would happen. But the general notion of a refinement is not entirely satisfactory.

Refinements and counter-theoreticals

Consider the game depicted in Figure 5.5. The simplest of refinements, backward induction, suggests that in this game the 'solution' is that player A begins by choosing D, anticipating that B would choose r (given the chance) in the further anticipation that A will choose R'.

But now put yourself in the position of player B and suppose that contrary to this theory player A chooses R. You now have the very best kind of evidence that player A is not playing according to the theories you hold; player A didn't do the theoretically anticipated thing. So are you now so confident that player A will conform to the theory and choose R' if you (as B) choose r? After all, you will suffer mightily if your hypothesis that A will choose R' is wrong.

The game in Figure 5.5 is drawn so that any answer except for a resounding 'yes' to the last question draws us into a quagmire. If one can anticipate that player B will choose d if faced with the counter-theoretical choice of R by A, then R by A becomes A's best choice. But then if B believes that A chose R at the outset in the hope that B would be scared into d, perhaps B should be relatively more confident that A will in fact choose R' given the opportunity and so r is safe for B after all. But in this case, shouldn't A have refrained from R?

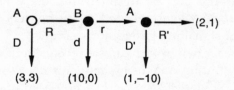

Fig. 5.5.   An extensive form game illustrating
the problems of counter-theoretical actions

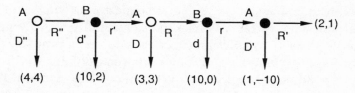

FIG. 5.6. More on counter-theoretical actions

And if you are not sure what to make of the predictions of backward induction in this case, adding another pair of nodes to get the game in Figure 5.6 should compound your confusion. Here A is supposed to begin with D″, at least according to the theory of backward induction. If she begins with R″, should B respond with r′, as theory dictates? And if he does, what conclusions should he draw when A subsequently (and counter-theoretically) chooses R?

Of course, the 'limit' of this compounding of difficulties takes us to games like the centipede game. But we reserve further discussion of that game for a little bit.

Let me insert two remarks about this and then sum up:

(1) Note that the consideration advanced here has relatively less impact on the von Stackelberg variation of entry deterrence because in that case each party moves only once; if one player catches the other in a counter-theoretical move, the first needn't worry about subsequent actions by the second. But in the variation depicted in Figure 4.7, we can be less comfortable; the theory suggests that the monopolist will charge the standard monopoly price in the first period, and the entrant is left to worry what it means if the monopolist does otherwise.

(2) The game in Figure 5.4(a), viewed as a strategic form game, is the game in Figure 5.7(a). We see quite clearly there the two Nash equilibria U-l and D-r, and we see that the first of these involves B choosing a weakly dominated strategy. That is, l does as well as r for B if A chooses U, but does strictly worse if A chooses D. Elimination of l on grounds that it is

weakly dominated by r is the strategic form 'equivalent' to the first stage of our roll-back procedure; in general, rolling back a game of complete and perfect information can be fully mimicked by iterated application of weak dominance in the corresponding strategic form. When discussing backward induction in extensive form games of complete and perfect information, however, authors (in particular, I) sometimes make the point that thinking about this argument in the extensive form strengthens the prescriptions of this refinement over what it would be if the game was thought of as a strategic form game. The argument for this runs as follows: If we think of this as a strategic form game, then perhaps it is possible for player B to be absolutely certain *ex ante* that A will choose U, in which case l is just as good as r. But if we think of this as an extensive form game, and if A chooses D, she puts B on the spot. B can no longer entertain beliefs that A won't choose D; A has chosen D, and B is left to respond as best he can. [12]

The consideration advanced here turns this 'strength' of extensive form thinking somewhat on its head. The strategic form equivalent of the game in Figure 5.5 is depicted in Figure

[12] Once again we touch on the question posed in Chapter 3, Is an extensive form game the same as the strategic form game to which it corresponds? If there is any sense to the argument just sketched, then there must be some sense in which the answer to this question is no. Kohlberg and Mertens (1986), who argue eloquently that the answer to this question must be yes, propose as the appropriate test '[Is there] an example of two game trees with the same strategic form, and whose "reasonable equilibria" are completely different— say every equilibrium payoff whatsoever is patently implausible in one of the two trees (p. 1012).' Since my interest in game theory is more an interest in predicting how individuals act, let me propose a different and easier test: Can we find a pair of extensive form games that give rise to the same strategic form such that, when played by a reasonable subject population, there is a statistically significant difference in how the games are played? And since that makes it too easy, if we find significant differences, can we organize the differences according to some principles that turn on recognizable differences in the extensive forms? For example (and following the discussion above), are we more likely to see play of D-R in the game of Figure 3.7(*d*) than in the games of 3.7(*b*) or 3.7(*c*)? To my knowledge these experiments have not been conducted, and the reader is free to conjecture what the results would be.

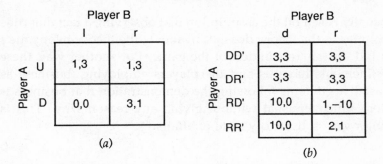

Fig. 5.7.   The strategic form games corresponding
to Figures 5.4(*a*) and 5.5, respectively

5.7(*b*). Corresponding to the backward-induction argument for
5.5 is the following chain: RD′ for player A is weakly dominated by RR′; once RD′ is eliminated, d for player B is weakly
dominated by r; once r is eliminated, RR′ for A is weakly dominated by the two (equivalent) strategies that begin with D.[13]
Thinking about this as a game in strategic form, B can reason
that as long as A assesses some probability that he (B) will
choose r, then RD′ cannot be optimal and hence r is better
than d. Similar logic drives us (and, presumably, B) to the
conclusion that A will choose D. But when we think of this as
a game in strategic form we do all our reasoning (and envision
all the players doing their reasoning) *ex ante*. Thus we never
put B in the uncomfortable position of having to reason *ex post*
about A's further intentions once A has chosen R, despite what
the theory suggests.

The main point is that we depend in many economic applications of game theory on refinements such as backwards
and forward induction (and more complex variations on these
themes) to select among Nash equilibria. Such refinements are

[13] The strategies DR′ and DD′ are equivalent for A because if A begins with
D, her subsequent plans to choose either R′ or D′ are irrelevant both to her
and to B no matter what B plans to do.

usually based on the assumption that observing a fact that runs counter to the theory doesn't invalidate the theory in anyone's mind for the 'remainder' of the game. Put another way, these refinements have us and the players employing 'business as usual' arguments following the demonstration that business is very much unusual without giving any reason why. This is not a very satisfactory state of affairs.

Complete theories

To get to a more satisfactory state of affairs we must answer the question, What does it mean if some player takes an action that the theory suggests he or she will not take? Why did it happen, and what does it portend for the future?

One of the responses in recent literature to the problem of countertheoretical actions is to try to see how answers to these questions affect our confidence in standard refinements. Authors pose 'complete theories'—theories in which no action or series of actions is absolutely precluded, although those that are not part of the equilibrium are held to be very unlikely a priori—and then draw conclusions for the refinements we commonly use.[14]

Of course, once we have a theory of why 'counter-theoretical actions' happen, those actions are no longer counter-theoretical at all, and so I will change to the term *counter-equilibrium actions*.

The seminal complete theory is not at all a recent innovation; it is Selten (1975), which with Selten (1965) began the literature on refinements. Selten advances the theory that any player at any information set might take any action by accident; when we see an action that runs counter to the equilibrium, we (and the players in the game) put this down to a slip of the hand that chooses the action. (Indeed, in the literature this is referred

---

[14] The term 'complete theories' and much of the discussion that follows has been developed in joint work with Drew Fudenberg (Fudenberg and Kreps 1990) and David Levine (Fudenberg, Kreps, and Levine 1988).

to as 'trembling-hand perfection'.) Such mistakes have very small probability; players usually carry out on their intended actions. Selten further postulates that in extensive form games a mistake made by a player at one information set does not in the least bit increase the chances of a mistake by that player or any other at any other information set. So, for example, in the game in Figure 5.5, if player A is observed to choose R, this is a mistake that doesn't affect the small chance (assessed, in particular, by B) of a second, subsequent mistake at A's second information set. Hence B can confidently choose r; A might err again, but as long as the odds of a second error are very small, r is better than d.

If this is our model of how counter-equilibrium actions are taken, then one sees the rationale for backward induction and similar refinements. What about forward induction (the argument for the game in Figure 5.4(*b*))? If in that game player A intends to play L, so that M and R are mistakes, then our argument against the L-l equilibrium seems to turn on an argument that one sort of mistake is more likely than another. We want to conclude that M is a more likely mistake than R. One can think of stories to support this; perhaps players guard more carefully against mistakes which hold out no prospect of doing better than following the equilibrium than they do against mistakes that hold out some (albeit small) chance of doing better. But equally sensible arguments on the other side are available. Myerson (1978) suggests that the likelihood of a mistake should be related to the cost of making the mistake; the more costly the mistake (given what others are meant to be doing), the more careful one is to avoid it, hence the less likely it should be. In the game in Figure 5.4(*b*), if we suppose that the equilibrium behaviour is L-r, then M by mistake gives A a payoff of −10 while R by mistake gives A a payoff of 1. M is a much more severe mistake and so, Myerson concludes, M is much less likely than R. But if M is a less likely mistake than is R, and if B views the choice of 'not L' as the manifestation of a mistake by A, then 'not L' leads B to conclude that M is less likely to have been A's choice than is R, which in turn leads B to choose r, which supports the L-r equilibrium.[15]

Whereas refinements of the backward-induction variety are justified by Selten's notion of hands that tremble statistically independently at various information sets, other stories about why counter-equilibrium actions are observed can cast doubt on such refinements. One such story, which is particularly damning of the standard refinements, turns on the idea that players may be uncertain about the payoffs of their rivals. To explore this story a brief digression is required.

What are a player's payoffs?

When economists build a model and write down payoffs, those payoffs are usually related to the 'pecuniary' incentives of the players. If the range of pecuniary payoffs is large, we can put risk aversion into our analysis with a von Neumann–Morgenstern utility function, but when physical outcomes are monetary, we usually assume that players prefer more money to less no matter how the money is obtained.

In every population of players, however, are individuals who respond to things other than money. In our discussion of the centipede game in Chapter 4, we invoked an individual whose (presumably) high moral character did not allow him to play d to the detriment of his fellow player. Similarly, we imagine that there are players whose low moral character might cause them to play in order to spite their rivals, even at some personal monetary cost; others who have an irrational aversion to playing strategies that have certain labels; and so on.

To take a concrete example, imagine the following bargaining game. There are two players, A and B. A can propose to B any split of $10, so much for herself and so much for B. B

---

[15] For the sophisticated reader, think about how Myerson's criterion, which is called *properness*, cuts when B plays the randomized strategy l with probability 3/14 and r with probability 11/14.

can accept or reject this offer, and if B rejects it, he can propose any split of $1. If the game gets this far, A can accept or reject B's proposal; if she rejects it, then each player gets nothing. This describes a game of complete and perfect information, and so we can solve it by backward induction. If B rejects A's offer, then B can offer one penny to A, which A will take in preference to disagreeing and getting nothing. Hence B can be sure of 99c if he is making the offer. But B cannot possibly get more than $1.00 if he is making the offer, so if A offers him, say, $1.01 on the first round, B is sure to take that. Hence A should offer B $1.01 (or, depending on how you resolve the ties, $1.00 or 99c), leaving herself with $8.99 at least.

The trouble with this analysis is that when this game has been played in the laboratory, players put in the position of B who are offered $1.01 or even $1.50 spurn A's offer. Note carefully, if B is offered $1.50 by A, then there is no way B can do monetarily as well rejecting as by accepting A's offer; B can get $1.00 at most by moving to the second round and he probably won't get A to agree to that. Indeed, what is seen in the laboratory is that Bs sometimes reject $1.50 and turn around with offers of 50c for each of the two.

The experiments that give these results are summarized and extended in Ochs and Roth (1989), who offer the following common-sense explanation: Player B is not interested solely in maximizing his monetary outcome. Within his tastes is a component that would feel like a dupe for letting A get away with $8.50 while he only gets $1.50. His taste not to be a dupe is strong enough to overcome his taste for the dollar or so he gives up by refusing $1.50 and proposing instead to take 50c.[16]

---

[16] What do you suppose would be the outcome of the following two experiments? First, suppose that prizes are multiplied by 100. Would as many Bs give up $100 or so in order not to feel like a dupe? Second, suppose that the initial proposal comes not from A but from some third, uninvolved party. That is, suppose the referee proposes a 'fair' split (where the referee judges what is fair), which B can reject and then make a counter-proposal. Would any B reject $1.50 if the referee proposed this initially?

His payoffs, therefore, are not simply given by the amount of money he receives or by any monotone function of that amount of money; to model this situation with a game that supposes that such a simple model is correct will lead to bad predictions.

Note well, not every B in these experiments rejects $1.50. Thus A's problem when formulating her initial offer is one of trying to assess with what probabilities her rival will reject offers of $1.50, $2.00, and so on. Suppose we simplify the game enormously, allowing A to propose either a split of $5 for each or an $8.50–$1.50 split. B must take the first if offered, but can turn down the second and then accept a 50c apiece outcome. Then if we identify the payoffs of the players with their monetary prizes, we get the game depicted in Figure 5.8(*a*). Backward induction tells us to predict that A will propose $8.50 for herself, and B will agree. But this is *not* a very good description of the situation. The situation instead is one in which B's tastes might run to turning down $1.50 and settling for 50c so as not to feel bad that A has profited from greed. We might, in consequence, model the situation as in Figure 5.8(*b*). Nature moves first, determining whether B is interested in money only or will turn down an inequitable offer by A. Note that we give the probability that B will turn down this offer as p. And then the game is played, but with A not knowing B's 'tastes'. In the top half of the game, where B is in it only for the money, we model her payoffs as being in terms of dollars only. In the bottom half, B's payoffs are changed. The precise numbers are unimportant; what is important is that B's payoff for spurning $1.50 and accepting 50c exceeds his payoff from doing otherwise. If this is the game in A's eyes, then A's determination what to propose depends on her assessment of p. If p is near zero, she asks for $8.50; if p is near one, she takes $5.00; and for some intermediate value she is indifferent.

Of course, the bargaining game originally described is much more complex than this. A has many possible initial offers,

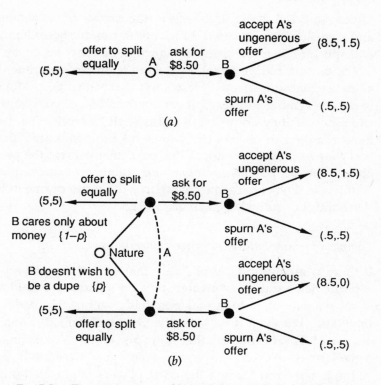

Fɪɢ. 5.8. Two caricatures of bargaining, with and without uncertainty about the motivations of one of the players

and B could be aiming to maximize his monetary payoff only, or he could spurn offers that leave him worse than $1.00, or worse than $1.50, or worse than $2.00, and so on. And B must worry about how A will respond to B's subsequent offer, especially as B will have declined A's offer. Will A try to exact revenge? And, then, if B is willing to spurn $1.00 (and let A have $9.00) if B can have 50c instead, is he still willing to do this if he predicts that A will reject his offer of 50c? Once we begin to introduce this sort of uncertainty into the game model, bargaining becomes tremendously complex, a thought we will return to at the end of this section.

But, to bring discussion back to the matter of refinements and complete theories, two things are important here. First, B will spurn \$1.50 in the game in Figure 5.8(*a*) if her tastes are as in the bottom half of 5.8(*b*), and there is ample experimental evidence that some people's tastes run in this direction. Money is not everything to every player, and models of competitive situations, if they are to be useful models of reality (i.e. predict behaviour in real contexts) must take this into account as best they can. And second, in this particular context the probability that B's payoffs are other than money-maximizing are significant. They aren't one chance in a thousand or one in ten thousand, but something like 0.4 or 0.25.

Complete theories built out of payoff uncertainty

If there is a substantial probability that one of the players in a particular game is motivated by non-monetary considerations, we should include that possibility within the model of the game. But even if we (and the players involved) think *ex ante* that it is very unlikely that some player plays according to a given set of objectives or payoffs, when a counter-equilibrium action is observed we and the other players might well put it down as evidence that the given player is so motivated. And then we and they may be led to assess relatively high probability for subsequent actions by this player that are 'crazy'.

If this is our theory of what manifests counter-equilibrium actions, then some backward-induction arguments can seem dubious. Consider the game depicted in Figure 5.9, thinking of the payoffs as being in pennies. Backward induction leads to the conclusion that A will choose RR' and B will choose r. But consider the possibility that A chooses D and B chooses d. (I purposefully haven't said what A's second action would be.) A's choice of D is perfectly sensible given B's intended choice of d. But in the spirit of backward induction we would say that B's choice of d is not credible; given the opportunity, A will finish with R', and so r is better for B than d.

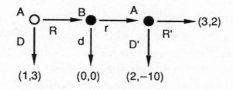

Fɪɢ. 5.9. An extensive form game illustrating the tension between backward induction and the slight possibility of malicious players

Player B might respond to this with the following argument: 'I expect A to play D, because A expects me to choose d. If A chooses R, then A is acting contrary to my equilibrium expectations, and my best guess is that A is doing this because she doesn't like me; she wishes me to get a small payoff, and she is willing to sacrifice her own interests to accomplish this. At first I didn't think there was much chance that A had such malice, but having seen R, my *ex post* assessment is that A is a nasty person. Which certainly justifies my choice of d. (And which in turn justifies the hypothesis that R by A is evidence that A is not playing to maximize her payoff.)'

This argument is not meant to convince you that A should play D; I'm not claiming that D-d is the obvious way to play this game. But at the same time, if when counter-equilibrium actions occur players reason in this fashion, the case for RR'-r as the obvious way to play this game (based on the logic of backward induction) is not so clear.

One can pose the question generally. Which equilibria (or even non-equilibrium prescriptions of action) can be definitely precluded if one attributes counter-equilibrium observations to the possibility that the player so acting has radically different payoffs from those considered in the model? We mean to consider situations in which players assess small probability *ex ante* that others have radically different payoffs, but where those small *ex ante* probabilities might loom quite large following a counter-equilibrium action. Formal answers to this question can be found in the literature, answers that turn on the precise way the question is posed.[17] But the rough spirit of the answers given is that at best one cannot follow backward-induction prescriptions for more than one iteration for each individual player. Refinements that are based on more than this are 'delicate'; they do not withstand explanations of counter-equilibrium observations that follow the sort of complete theory that we have outlined.

The delicacy of equilibrium

Carrying on with this logic a step further, we can ask whether our focus on Nash equilibria is appropriate when counter-equilibrium actions are explained by complete theories of this sort. That is, in an environment where counter-equilibrium actions indicate that a player may have payoffs radically different from those in the model, a possibility of small probability a priori, is it still the case that 'self-evident ways to behave' must pass the Nash equilibrium test?

As noted in Chapter 4, for all practical purposes the answer is no.[18] In the centipede game and games like it, where there is a large number of moves back and forth between or among the players, a small probability *ex ante* that players have radically different payoffs from those in the model can completely reverse the predictions of Nash equilibrium analysis. If in the centipede game there is a one in ten thousand chance that your rival is the soul of co-operation, or if your rival assesses this about you, then it becomes self-evident (after a fashion) to choose R instead of D at the outset of the game, something

[17] See, e.g., Fudenberg, Kreps, and Levine (1988).

[18] There is a theoretical sense in which the answer is yes. Fixing a game, if the *ex ante* probability that a player has radically different payoffs is made extremely small, then all the 'self-evident ways to behave' with the possibility of radically different payoffs included in the model will resemble Nash equilibria of the game modelled without that possibility. But, for example, in the centipede game, the 'extremely small' probability must be on the order of $(0.5)^{100}$; reasonable but still small probabilities of radically different payoffs— say 0.0001—are very far from small enough.

that is part of no Nash equilibrium if this sort of possibility is not modelled.

In Chapter 4 I said that the observation that this is so is a success of the theory; here I seem to be listing similar observations (about the delicacy of the refinements) as a weakness. Let me be clear why this is.

Any formal, mathematical theory is applied to a model of a given situation. This model will at best be an approximation to the situation. It is therefore crucial to have a sense—intuitive if need be but better if buttressed by theory—of *when* the conclusions of the theory turn on features of the model about which the model-builder is fairly uncertain. In such cases one trusts the conclusions of the theory at great peril. The theory is trustworthy in cases where its conclusions are not reversed or greatly altered by changes in the formulation that seem 'small' intuitively; to know when the theory can be trusted, it is useful (to say the least) to know when it cannot be.

Hence we can read the conclusions drawn in this section as a success because they teach us about the limits of applicability of some tools we commonly employ. If counter-equilibrium actions are explicable by this sort of complete theory, then predictions based on backward induction (and extensions of backward induction) may be untrustworthy past one 'roll-back' per player. Unhappily, many of the apparent successes of game theory in economics are based on the application of multiple rounds of backward induction. It is good to know that all these applications should be regarded with scepticism. But this insight does not lead to the conclusion that many apparent successes of the theory in modelling economic phenomena are necessarily real.

Alternating-offer bargaining

Both because we have discussed bilateral bargaining at some length in this chapter and because it illustrates some of the points made above, let me close this section by discussing a very important recent innovation in bilateral bargaining, models that use alternating-offer protocols.

Consider the following variation on the bilateral bargaining stories from earlier in the chapter. The two parties involved must agree how to split a sum of money. Any split of the money is feasible and if the two come to no agreement then each gets zero. What is a bit different in this formulation of the problem is that the 'pie' that they have to split shrinks as time passes. If they can agree on a division of the money at time $t = 0$, then they can split up to \$100. But at time $t > 0$, the amount of money that they can split is only $\$100\delta^t$ for some given number δ between zero and one. To fix matters, suppose time is measured in hours, so $t = 2$ is two hours from the start of negotiations, $t = 1.5$ is an hour and a half, and so on, and suppose $\delta = 0.5$; the 'pie' shrinks by half each hour.[19]

The fact that the pie is shrinking does not change the basic conclusions that we came to concerning simultaneous-demand bargaining. Suppose, for example, that bargaining is conducted as follows: Every thirty seconds, beginning at $t = 0$, the two parties simultaneously and independently announce percentages of the total (remaining) that they wish. The first time their offers sum to 100 per cent or less, the clock stops and they split what remains in the fashion they demanded. If, for example, they come to agreement on the seventh round (after 3 minutes = 0.05 hour), agreeing to 60 per cent to player A and 35 per cent to player B, then A gets $(0.6)(\$100)(0.5^{0.05}) = \$57.9562$ and B gets $(0.35)(\$100)(0.5^{0.05}) = \$33.8078$. In this situation, any split of the \$100 on the first round of bargaining is a Nash equilibrium. And more besides; there are equilibria in which agreement is only reached after, say, ten minutes have elapsed, and the pie has shrunk appreciably.[20]

---

[19] In less stylized formulations of the problem, δ is related to the time-rate of discounting that the two parties apply to the fruits of any settlement they might reach.

[20] Can you construct such a Nash equilibrium? The key is that a player demands more from his rival if that rival attempts to settle 'too soon'.

But now change the rules a bit. Suppose that only one of the two players is able to make an offer at a given time. You can think, if you wish, of a situation in which the players have a large stack of blank contracts of the sort in Figure 5.1(*b*); player A fills out a contract at the start of the game and gives it to player B to sign. If player B signs, that terminates negotiations in agreement; if not, then player B fills out a contract and gives it to player A for her approval. If player A agrees, negotiations terminate in agreement; if not, it is A's turn once again to make a proposal, and so on. That is, the players alternate in making offers, which gives this bargaining protocol the name *alternating-offer bargaining*.[21] Imagine that it takes each player exactly fifteen seconds to fill out a contract. Approval on the other hand is instantaneous.

It still seems that virtually any split of the money is an equilibrium. After all, if player B adopts the strategy of refusing any offer that leaves him with less than 80 per cent of the pie and always offering 20 per cent to A, then A's best response is to offer B 80 per cent at the outset. And if A's strategy is to offer 80 per cent each time and refuse less than 20 per cent, among B's best responses is the strategy outlined above.

But while this is a Nash equilibrium, it turns out that it is based on an incredible threat by B. Suppose it is A's turn to offer and she offers B only 79.9 per cent. B is meant to turn this down and insist next time on 80 per cent. But to do this costs B a little bit in terms of the fifteen-second delay; if the pie is x when A makes her offer (so B stands to get $0.799x$ if he agrees), even if A agrees to an 80–20 split next time, B will net only $(0.8)(0.5^{1/240})x = \$0.7977x$. Why should B turn down $0.799x$ to get only $0.7977x$?

From extensions of this argument, Rubinstein (1982) derived the following remarkable result. While the alternating-offer

---

[21] The story of bargaining connected to Figure 5.8 is an example of alternating-offer bargaining but with a very quickly shrinking pie and only one round of offer and counter-offer.

bargaining game described above has many Nash equilibria, it has a single equilibrium that does not involve any incredible threats of any sort, one in which the first player to make an offer offers to take a little bit more than 50 per cent of the pie, and the second player agrees immediately.[22]

What do we mean by 'an equilibrium that does not involve any incredible threat'? The game we have described is one of complete and perfect information, but it is also one with an infinite game tree (there is never a last offer) and so we cannot apply backward induction. Rubinstein's original argument is based on invocation of the refinement known as *subgame perfection*, but there are other arguments coming to the same conclusion. In particular, if you think of the game as lasting, say, no more than thirty days (so the original $100 shrinks to ($100)($0.5^{(24)(30)}$) = 1.813×10^{-215}), then you can apply backward induction to the now finite game of complete and perfect information, and you will get the answer given above (to a reasonable degree of approximation).[23]

Before concluding that we now know how to get a definite solution to the bilateral bargaining problem, let me suggest three variations.

(1) Players act just as in the protocol above except that A happens to be blessed with better reflexes; it takes A five seconds (only) to fill out a contract, while it takes B fifteen seconds to

---

[22] For the parameterization given here, the first player takes approximately 50.07222 per cent. The exact formula and details of the proof of this result can be found in Rubinstein's seminal paper, or see Kreps (1990) or Osborne and Rubinstein (1990).

[23] (a) You will need to know how to handle ties in backward induction if you do this, since they are endemic to this story. Also, the analysis of alternating-offer bargaining with a finite horizon predates Rubinstein and is due to Stahl (1972). (b) The sense in which the infinite move alternating-offer bargaining model is approximated by models with a finite but distant horizon is a bit more complicated than this throw-away assertion may indicate. What makes the approximation valid is that we get approximately the same answer no matter what division of the remaining 1.813×10^{-215} is enforced if an agreement is not reached within thirty days.

do so. Then there is a single Nash equilibrium that is not based on incredible threats, which gives approximately $75 to player A and $25 to B. That is, the fact that B takes fifteen seconds to fill out a contract and A five seconds means that A gets three times as much as B.

(2) Suppose that the situation is precisely as in the original formulation (it takes each player fifteen seconds to make a proposal), but proposals must be in multiples of a $1.00. Then every split of the $100 is once again a completely credible equilibrium. (If the pie shrunk to half every two days instead of every hour, then the same would be true if offers must be in multiples of a penny.) Rubinstein's result depends crucially on our ability to divide money finely.

(3) Suppose that the pie doesn't shrink at all; there is always $100 to split. But the players must provide their own contract forms. It costs A 1c every time she makes a proposal to B, and it costs B 1.1c every time he makes a proposal to A. Then the game has a unique 'credible' equilibrium in which A asks for and gets $100 if she makes the first offer; if she makes the second offer, she is offered and settles for $99.99. If it costs one player a tiny bit more than another to make an offer, then the first player gets almost nothing and the second almost everything.

One of the supposed virtues of non-cooperative game theory is that it allows us to investigate how changing the 'rules' of a particular competitive interaction changes the outcomes we expect. We are seeing that rules affect outcomes in this litany of examples, but surely these drastic changes are not intuitive. Also, we earlier expressed the hope that the outcomes of bilateral bargaining would not be too sensitive to the exact bargaining protocol employed, so that we could come to some prediction about what happens when we lock the two parties in a room and let them negotiate as they wish. In view of this litany, our earlier hopes seem rather forlorn.

However (coming to my point in telling you all this, besides providing a glimpse of some of the nicest work that was done in game theory in the last decade), all these games share the back-and-forth, relatively-little-ever-at-stake characteristics of the centipede game. In any of these variations on the rules, if we included in the model the possibility that player A suspects that player B might irrationally insist on 90 per cent for himself, even if A gives this possibility only a 0.01 chance a priori, then the variations in the rules are overwhelmed; B will indeed get (at least) 90 per cent or so of the pie in any credible equilibrium.[24] What happens when A has more complex suspicions about B and B entertains similar suspicions about A? To the best of my knowledge this is a question that has not been adequately addressed yet, although I suspect that the answer, when it comes, will continue to indicate that seemingly small variations in the bargaining protocol will be swamped by this sort of uncertainty by each player about the 'objectives' of the other. Or, to be more precise, *small* variations in the bargaining protocol will be swamped in cases where there is not much delay between or other costs of offers and counter-offers, whatever is the exact protocol. If it takes players thirty minutes or so to put together offers and the pie shrinks by half every hour, then variations in the bargaining protocol become 'serious' and (at least by my intuition) should continue to matter; it remains to see whether this intuition will be confirmed by analysis.

The rules of the game

As we have noted all along and seen especially starkly in the examples just given, the 'rules of the game' can matter enormously to the outcomes predicted by game-theoretic analysis.

---

[24] So I do not get myself into too much trouble here, let me make the last assertion into a conjecture; I don't think a proof has ever been written down. Of course, this conjecture presumes that no other changes are made to the original formulation.

This brings the question, Where do the rules come from? And the implicit logic of causation is that changing the rules of the game can change the outcomes; rules are given exogenously, and they then affect the outcomes. But is causation one way only? Can outcomes affect the rules?

For the most part, game-theoretic analyses take the rules as given and proceed from there. There have been some analyses in which, in early stages of 'game', players choose rules that will be in force in later stages. Analyses of the entry-deterring monopolist in which the monopolist in the first period takes concrete actions that affect the conditions of second period competition can be viewed in this fashion. Another example in this spirit is the recent paper by Klemperer and Meyer (1989) in which duopolists choose whether to be price-setters or quantity-setters (or something between). But despite such analyses, I think it safe to say that *game-theoretic analyses in economics tend to take the rules of the game too much for granted, without asking where the rules come from. And they do not consider very well whether the rules that prevail are influenced by outcomes.*

Let me illustrate what I mean with an example documented by Sultan (1975) and Porter (1983).[25] In the United States in the early 1960s, two firms manufactured large electric turbine generators, General Electric and Westinghouse. A large electric turbine generator is a machine that produces electricity from mechanical energy (provided by steam from a nuclear, coal-, oil-, or gas-fired plant, or by running water). These machines are very large and expensive, and there are large fixed costs associated in their manufacture (especially, R&D costs), so these two firms were relatively confident that (except for the possibility of foreign competition), they did not face the threat of entry. Large electric turbine generators were produced to order for electric utility companies by the two firms; demand for these machines was highly cyclical (an increase in the demand

---

[25] The description that follows is highly simplified, and you would do well to read about this case in all its complexities.

for electricity in one part of the US would often be correlated with an increase in demand for electricity elsewhere); and orders were frequently backlogged.

On the face of it, this seems like an ideal industry for effective implicit cartelization in the sense of Chapter 4; both GE and Westinghouse would charge high prices, under the threat that either would revert to very low prices if the other tried to undercut the first. But in fact profits were not at all high when this story opens, and a game-theoretic analysis indicates why. When a large utility decided to buy a turbine generator, it would first negotiate with one manufacturer, then with the other, then back to the first, and so on. These negotiations were conducted behind closed doors and involved not only the price of the generator but also its exact specifications, the supply of spare parts, the delivery time, and so on. If GE maintained a high price and then lost the order, it did not have the information needed to conclude that Westinghouse had violated an implicit cartel agreement, because it did not know just what Westinghouse had offered. And, as the theory tells us, when one attempts implicit collusion in a 'noisy' environment, where parties to the collusive scheme cannot see very well what their rivals are doing, then collusion is difficult to maintain.

So GE, in rather dramatic fashion, decided to change the rules. It put in place a number of institutions which made it very easy for Westinghouse to know how GE would bid in any particular instance. In particular, it offered its customers 'price protection'; if GE lowered its prices relative to a unambiguous book-price for one customer, then it owed similar discounts to all customers who had bought from it over the previous few months. The effect of this was two fold. GE was contractually obligated to reveal the prices it was charging; it even employed a public accounting firm to make those revelations credible. And it became more expensive for GE to cut prices for one customer to land a particular sale, thus assuring Westinghouse that GE wouldn't shade its prices cavalierly.

After a bit of confusion, Westinghouse adopted very similar measures. Now there was little problem for one firm to know *ex ante* how the other would bid, and if one firm lost a sale, it knew *ex post* precisely what the other had done. Implicit collusion became quite easy to maintain, and maintain it the two firms did, reaping substantial profits until the US government stepped in and put a halt to the institutions (such as 'price protection') that made the collusion feasible.

This is a case that is often taught when teaching about game-theoretic approaches to implicit collusion, because it so cleanly makes the point that implicit collusion is difficult in noisy environments and easy in less noisy ones. Budding oligopolists (e.g. MBA students at Stanford University) draw the obvious normative conclusion; if you are in an oligopoly situation in which collusion cannot be sustained, and if you can't somehow turn this into a monopoly—always the first course of action favoured by budding oligopolists who would much rather be budding monopolists—then look for ways to change the situation so that collusion becomes feasible.

But this begs the question, Why did GE change the 'rules' when they did? Why didn't they change the rules before? Did the rule change *per se* simply allow the two firms to fall into a collusive scheme (which is what the theory suggests)? Or was it the fact that the rules were being changed by one of the participants that caused the subsequent collusion? If we analyse the situation first with one set of rules—concluding that collusion will be difficult—and then with a second—concluding that now it will be possible—we have missed out on a very important part of the dynamics of competition in this industry, exactly what game theory is supposed to help us understand. As I noted, we can try (as have a few authors) to incorporate 'rule changes' into our formulation of the game. But this is not much done in the literature; and in any event I do not believe that conscious profit- or utility-maximizing choices by individuals will give us very good explanations of how the rules came to be what they are and how they will evolve. To explain this last remark, we must move on to my proposals for the further development of the theory.

6
Bounded rationality and retrospection

To summarize where we are, non-cooperative game theory has been of value to economists because it has given us a language for modelling dynamic competitive interactions and because it allows us to begin with an intuitive insight in one context and build out in two directions, either taking the intuition from one context to another or extending and probing intuition in somewhat more complex settings. It has not been especially useful when applied formally to contexts that are too far removed from its intuitive base, and it has not been successful in explaining or organizing the sources of intuition. In particular, it has left relatively untouched the following fundamental questions:

(1) In what contexts is equilibrium analysis appropriate?

(2) If equilibrium analysis is inappropriate, what (useful) notions can we use instead?

(3) In cases where equilibrium analysis is appropriate, why is it appropriate?

(4) In cases where equilibrium analysis is appropriate and there are multiple equilibria, which equilibrium should be selected? (And, in particular, what should we, and the players involved, make of counter-equilibrium actions?)

(5) Since game-theoretic techniques require and, indeed, trade heavily upon exogenously given rules of the game, where do specific rules come from, how do they evolve and change, and what do we make of relatively free-form competitive interactions?

In this final chapter, I wish to suggest an approach to these questions. This approach will not resolve all these questions completely, and it may not resolve some of them at all. But this approach gets at some of the roots of these questions and so leads us towards a resolution of them, or so I will attempt to convince you.

The expositional flow of this chapter is somewhat round-about. I begin with a fairly long prelude, returning to a question mentioned in Chapter 3 and that was, at least implicitly, the focus of the second part of Chapter 5, viz. Why do we study Nash equilibria? There are two reasons why I focus on the concept of Nash equilibrium: The question is important in its own right; in many of the recent applications of game theory to economics, equilibrium analysis is the methodology employed.[1] More to the point of my current purposes, this specific question leads fairly directly to models of individual behaviour that is boundedly rational and retrospective. None the less, this prelude is lengthy, and it is only when we get (in the second section) to models of boundedly rational and retrospective behaviour that the main point of this chapter will be reached. So the reader who prefers a somewhat more coherent organization might wish to regard the first section as a short chapter in its own right that elaborates issues raised earlier, with 'Bounded Rationality and Retrospection' beginning there-after.

Why study Nash equilibria?

Why do we study Nash equilibria? Recall from Chapter 3 the answer I began to develop. If the game in question has or will come to have an evident way to play, then a necessary condition for that evident way to play is that it is a Nash equilibrium.

---

[1] Do not prejudge the issue, however. We may well conclude, along with a number of authors, that the emphasis on Nash equilibrium analysis is mis-placed. Discussion of this point follows.

Games where the set of Nash equilibria is not relevant

You may also recall from Chapter 3 the statement that when a particular game has no evident way to play and we do not believe that one will be developed (say, via negotiation), then there is no reason to study the set of equilibria. To begin to understand why the set of equilibria might be relevant, let me give a few examples where it is not.

(1) *A game with a unique Nash equilibria that is rarely played.* There are two players, A and B. The game has either one or two stages. In the first stage, A and B simultaneously and independently choose between the letters X and Y. If both choose Y, both receive $1 and the game is over. If one chooses X and the other Y, both receive $0 and the game is over. If both choose X, the game proceeds to a second stage. In this second stage, the two simultaneously and independently choose positive integers. These integers are then compared and pay-offs are made; if the integers chosen are different, the player choosing the higher integer wins $250 and the player choosing the lower integer wins $100. If the integers are the same, each player wins $25.

I have not staged this game because the stakes are more than my bank account can stand, but I have some casual evidence about similar games with lower stakes. That evidence suggests the following fairly obvious conjecture: If this game is staged between reasonably intelligent players, the first stage will conclude with both players choosing X. No conjecture is offered about the resulting second stage; that is a mess with no self-evident way to play. But we can be fairly confident that X will be chosen by each player in stage one.

The point is that this game has a unique Nash equilibrium in which each player chooses Y in the first (and hence only) stage. Proving that this is so is a bit more technical than I wish to be here, but the idea is relatively simple. If the players choose X with positive probability in some equilibrium, then there must be some equilibrium entailed for the 'name the greatest integer game'. But that game is not mathematically well behaved and has no Nash equilibrium at all.[2] Hence we have here a game in which there is a unique Nash equilibrium, and (I feel confident in predicting) we will rarely see that equilibrium chosen.

Of course, this is a game without an evident way to play. It is fairly evident (to most players) that X should be the choice in the first stage, but it is not at all clear what to do in the second stage. So, as claimed above, when there is not an evident way to play a particular game—a way to play the *entire* game—then observed behaviour may have nothing to do with the set of Nash equilibria.

You may conclude, and with some justice, that this game and games like it are too artificial to take seriously. But it does establish the principle; now the question is, Can we find more reasonable games in which the principle seems to apply?

(2) *Chess.* We have already remarked that chess is a finite game of complete and perfect information, so in principle we can use backward induction to solve it. Since it is also a constant-sum game, ties encountered while employing backward induction will not be a problem (if a player is indifferent between two or more possible courses of action, his opponent will be also), and all Nash equilibria of the game if there is more than one will give the same payoffs as the solution obtained from backward induction.

All of which, of course, is completely irrelevant. The game is too complex for backward induction, and while we might be interested in the set of Nash equilibria if ever we could get our hands on it, we can't do so, and so there is no (practical) interest in the equilibria.

(3) *A variation on the cities game.* Recall the cities game from Chapter 3. In particular, recall the variation in which each city

---

[2] For those who know about such things: it has no equilibria in pure strategies or in mixed. The strategy space for the second stage is simply too badly behaved.

was given a set number of points, and payoffs were: $1 per point if a city was on one list and not the other, to the player listing the city; $2 per point taken from each player if a city appears on both lists; double winnings if the players partition the list of cities. But now suppose the list of cities is changed to Caceres, Canoas, Carmocim, Carvoerio, Compinas, Cuiaba, and Curitiba, with Caceres mandated to be on the first list and Carmocim on the second. The set of Nash equilibria (in pure strategies) for this game consists of all partitions of the seven cities in two with Caceres and Carmocim separated. And if you wish to predict behaviour in this game for two players chosen at random out of, say, the audience of a talk given at Oxford University, I doubt you would bet much money on the possibility that they would attain one of these equilibria. No evident way to play, so no special interest in the set of equilibria (at least, for purposes of predicting the outcome of play by two randomly selected individuals).

(4) *A particular bargaining game.* Consider the following bargaining game. There are two players, A and B, who simultaneously and independently make demands for some number of 'poker chips' between zero and one hundred. If their two demands sum to one hundred or less, each gets the number of chips he demanded. If the two demands sum to more than one hundred, each gets nothing. Assuming the first condition holds, each player is then given a chance to win a prize, where the probability that a player wins his prize is equal (in percentage terms) to the number of poker chips the player holds. So, for example, if the players divide the poker chips 60 to A and 35 to B, than A has a 60 per cent chance of winning her prize and B a 35 per cent chance of winning his. A's prize is $40, whereas B's is $10. All these rules are made clear to the two players before they formulate their demands; each is aware that the other knows all this, and so on.

If the two players had identical prizes, say $40 each, then we might predict that in a game this symmetric, each player would confidently demand 50 chips, expecting the other to make a similar demand. That is, the symmetry of the situation, if it is symmetric, may suggest to the players a particular Nash equilibrium. But in this case where A's prize is $40 and B's is $10, matters are less clear. Should the two split the chips equally? Should A get all 100 chips (since she has the larger prize)? Should the chips be split, say, 20 for A and 80 for B, so that A and B both have an expected prize of ($40)(0.2) = ($10)(0.8) = $8? It isn't clear a priori which if any of these is 'appropriate', and this ambiguity implies that if we imagine the two players are chosen at random from a large population and are given no opportunity to discuss matters before bargaining, their play may not conform to any Nash equilibrium.

(5) *The centipede game.* Recall the centipede game (Figure 4.12). There is no need for a detailed rehashing of comments made earlier about this game. The predictions of Nash equilibrium analysis (that player A begins with D) are not empirically valid, and a likely explanation is that the predictions of Nash equilibrium analysis are very precariously set upon the assumptions that each player knows the payoffs of the other, each knows this, and so on. Very small perturbations in these assumptions, modelled formally, lead to very different formal predictions; so we cannot have much faith in the formal dictates of equilibrium analysis in such a case.

(6) *How to compete following deregulation of the US domestic air-transport industry.* Recall also the story of the US domestic air-transport industry, briefly mentioned in Chapter 5. In this fairly complex situation, players (rival firms) were unclear on what others would do, how they would behave, and what were their motivations. Each individual firm could try to make assessments of what others would do and choose accordingly optimal responses, but it would have been rather a surprise if the behaviour so engendered resembled the equilibrium of any game-theoretic model that didn't begin with that behaviour and then construct the model around it.

Reasons why there might be a self-evident way to play, or confident conjectures about the actions of others

These examples illustrate a mixed bag of reasons why equilibrium analysis may fail to give good predictions of individual behaviour in some competitive situations. Equilibrium analysis is based formally on the presumptions that every player maximizes perfectly and completely against the strategies of his opponents, that the character of those opponents and of their strategies are perfectly known,[3] and that players are able to evaluate all their options. None of these conditions will be met in all respects in reality. The behaviour of individuals in economic contexts may approximate these assumptions in some contexts, but the approximation may be insufficient in others.

When these assumptions are laid out in this fashion, it may strain credibility that they could ever be even approximately satisfied. What, if anything, would give rise to circumstances that approximate these assumptions?

As is made clear by the example of chess, a first requirement is that the situation should be simple enough so that players are able to evaluate the options that they consider they have. There is a point of some subtlety lurking here. In almost any real competitive interaction, players will have many complex options, the consequences of which will not be clear to them. When economists build a game-theoretic model of the competitive situation, many of these options are ignored, which then may make the *model* simple enough so that we can reasonably imagine players evaluating all the options they are given in the *model*. If the model is to capture adequately the situation it purports to represent, it is of course necessary that options omitted from or simplified in the model are treated by

---

[3] Or, for games with incomplete information, any uncertainty in the mind of one player about another player's character is 'fully appreciated' by all the players in the game, and the strategy of the second player as a *function* of his uncertain character is known to all.

the participants in similar fashion. Hence I have formulated the first sentence of this paragraph as, '. . . players are able to evaluate the options that they *consider* they have'. Even with this qualification, chess is too complex a game to meet the requirement of relative simplicity; players may not consider all the moves that are feasible for them under the rules, but (until a simple end-game is reached) the number of moves they do consider is more than they can reasonably evaluate fully. But certain situations of competitive interaction in economic (and other) arenas may be regarded by the participants as being sufficiently simple to make the necessary evaluations.[4]

Beyond this, the conditions for Nash equilibrium are that players are relatively certain how their opponents will act.[5] We are back at last to the issue raised in Chapter 3. Why in some contexts will individuals see that some mode of behaviour (for their fellows and themselves) is probably the mode of behaviour that will be followed? What resolves *strategic uncertainty*? I gave some of the usually cited reasons briefly in Chapter 3; here I will repeat those reasons and develop another.

(1) Players may have *the opportunity to negotiate before play* of the 'game'. Consider, for example, the following implementation of bilateral bargaining. Two players will, after thirty minutes have passed, make simultaneous and independent monetary

---

[4] For those who play it, the game of Go may illustrate the point. Go, like chess, is a constant-sum and finite game of complete and perfect information, so in principle it is solvable. But, like chess, it is fantastically complex taken in its entirety, and the theoretical existence of a solution is of no practical relevance. Unlike in the case of chess, however, *pieces* of the game can be simple enough to be solvable. Go, roughly put, is a game of territory; one part of the board can be isolated from the rest and (if the part is small enough) then 'settled'—the way to play in that part of the board is evident to both players—while other pieces of the board remain too complex to be solved. Since Go is a constant-sum game the analogy to economic life is not perfect, but the analogy to life is closer in this respect than is the case of chess.

[5] At least, up to any uncertainty about their opponents' characters; and then any such uncertainty must be 'appreciated' by all concerned.

demands. If those demands taken together lie in the shaded region of Figure 5.1(*a*), then they will be met; otherwise each player will get $1. And for the next thirty minutes, the players are given the opportunity to exchange views on how they should play. We do not model this pre-play exchange of views; the 'game' is simply the simultaneous formulation of demands by the two. But we may predict that the demands the two formulate are more likely to form an equilibrium of the demand game because the two have the opportunity to exchange views beforehand. We may be less sure what equilibrium they will achieve (if they achieve one), but the chances that they will achieve some (pure strategy) equilibrium are probably higher in this case because they are able to negotiate. (Needless to say, this is an empirical proposition, which ought to be tested in the laboratory.)

(2) Players may have *relatively direct experience* playing the game in question or some other closely related game. Imagine playing the seven cities game variation described in the previous subsection under the following conditions. Two players are chosen at random from a large population (the lecture audience at Oxford, say); they play the game and their choices are revealed to the audience. Then two other players are chosen at random and they play, with their choices revealed. Then two more, and two more, and so on. (Assume that while the choices of the players are made public, the points per city, and hence the payoffs, are revealed only to players.) It seems reasonable to expect that, in such situations, some *modus vivendi* for co-ordination will arise. (I have no direct experience playing this particular game, but I have observed similar phenomena in the classroom.) If there were eight cities instead of seven, I will happily bet at even odds that this *modus vivendi* would involve four cities for each player although I have no particular idea which four cities (and my willingness to bet is not based on any particular specific information; I've never run this particular experiment, even casually).

Roth and Schoumaker (1983) report a classic experiment along these lines, involving the previously described problem of bargaining over one hundred poker chips where each chip represents a 1 per cent chance at winning a prize, and the two players have different prizes. I will not go into details, but the rough spirit of their experiments runs as follows.[6] In earlier work, Roth and associates had noted a tendency for one of two 'equilibria' to appear; the 50–50 split, which equalizes chances of winning, and the 20–80 split, which equalizes expected monetary prizes. Subjects in this experiment bargained repeatedly, bargaining via a computer (CRT) interface. A subject either always had a $40 prize at stake or always a $10 prize. Unbeknown to the subjects, they were, in early rounds, bargaining against the computer, which was sometimes programmed to insist on a 50–50 split of the chips and sometimes on a 20–80 split. Subjects were 'conditioned' consistently; for any given subject, either the computer insisted generally on one split or the other. Then subjects were paired (via the computer), with the obvious results that the conditioning worked very well; players tended to stick to what they had learned as the evident way to bargain.[7]

In discussing direct experience as a justification of Nash equilibrium, we must be particularly careful about cases in which there is repeated interaction among a small group of players. When the same players or a small group interact repeatedly, then reputation and folk theorem-like constructions (as in Chapter 4) intrude, which complicates matters. In such a situation, by relatively direct experience do we mean repeated

[6] My summary emphasizes those aspects that are important to my thesis and suppresses some other aspects that are not. You should consult their paper for a fuller, more accurate explanation of what they show.

[7] The most interesting subjects, of course, were those whose conditioning was contradicted by subsequent experience; a $10-prize subject, for example, who had been conditioned to expect 80 chips and then was paired with a $40-prize subject who had been conditioned to expect 50 chips. So that you have incentives to search out this article, I leave it to you to conjecture and then read how interactions between these subjects progressed.

interactions among the particular small group of individuals or do we mean relatively direct general experiences with repeated interactions among small groups, of which the particular interaction is one case? Trying to have things both ways, we mean both. My ongoing relationship with my students, for example, is to some extent formed by my experiences in relationships I have had with other students (and they with other teachers) and by the experiences other professors have had with other students, as related to me and to my students. At the same time, my repeated interactions with student A teach each of us specific things about the other; you might think of our early interactions as a 'feeling out' or negotiating process in which our particular relationship is formed. NB: these are two potentially very different phenomena, and one should not necessarily expect theories of them to be similar.

(3) Closely related to direct experience is less direct experience. I split this category into two. First are *social conventions* (merging into social norms). An example would be the battle of the sexes game played by a Korean graduate student and his professor. Although I lack the data to make such an assertion, I strongly suspect that two such players would have little trouble seeing how to play; deference in one direction is clearly indicated. Another example would be bargaining over the hundred poker chips if the two prizes were precisely the same. I would expect in this case that the two settle on a 50–50 split; the rule of equal division should be fairly well ingrained in such cases, at least if the two subjects are peers. (What would a Korean professor and student demand in such a situation? Based on very limited experience, I can easily imagine the student asking for nothing; I have no experience whatsoever with Korean professors, so I have no idea what they would demand.)

(4) Moving along a continuum from situations in which social conventions clearly apply are situations in which individuals have a good sense, somehow socially or culturally derived,

what to do, but they would have a hard time articulating from where this sense comes. If we were to ask these individuals to explain the origins of this sense, we are likely to hear, 'Well, it's just obvious.'[8] The cities game with nine European capitals is a good example. Following Schelling (1960), whose discussion of these things is still the best of which I am aware, the term *focal-point equilibria* will be used for such cases.

(5) Finally, and seemingly quite different from the categories given above, is that players may be able to use *deduction and prospection* to see how to behave. A very simple example is given by the game in Figure 3.7(*d*), reproduced here in Figure 6.1. Player A thinks ahead. If she chooses D, then B must choose either −1 (L) or 0 (R). It seems clear that B will choose R. Hence A's choice is between payoffs of 1 (U) or 2 (D), and D is the better choice. That is, A deduces how B will act and chooses her response accordingly. Applications of dominance or iterated dominance and of forward induction would seem to fit here. Indeed, most of game theory is couched as an attempt to deduce how 'rational players' would act in a particular situation, which can be thought of in this light; a player reasons deductively about the prescriptions of 'rationality' in order to know how others will act, and then chooses his own best response in reply. There have been attempts to follow such a programme for all games, the most notable of which is Harsanyi and Selten (1988). But even if one doesn't think that this programme can be carried out universally, there are certainly some games—such as the game in Figure 6.1—for which this sort of reasoning works.

Thinking about these five stories, it would seem that (1) and (5) are different from each other and from the others, while (2), (3), and (4) are points along some sort of continuum. The boundaries between (2), (3), and (4) are vague at best. When does previous experience count as 'relatively direct'? When

---

[8] I used precisely this explanation, e.g. in Chapter 3.

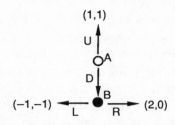

FIG. 6.1.   A game solved by deduction and prospection

does a social convention (Korean students defer to their professors) become a focal point (political arrangements are an obvious way to divide a particular set of cities)? It seems to me nearly impossible to say, especially since we have no formal constructs at hand (yet) to make a division.[9] For the time being, therefore, I am happy for you to think of (1) through (5) as three stories instead of five, where the middle story (about personal, social, and cultural experience) takes on many diverse forms.

Comparing the examples and the stories

We now have a number of examples of games without an evident way to play, hence in which the set of Nash equilibria is not of any special interest to us, and some stories about why games might have an evident way to play. It is instructive to take the examples in turn and see what it is they lack in terms of the stories.

(1) The game that can lead to 'name the greatest integer' is simply badly behaved. If we considered just the subgame in which players try to name a greatest integer (with prizes as in the original game), this game has no Nash equilibria at all, and it cannot (therefore) have an evident way to play. Some

---

[9] I will hint at some formal constructs later this chapter, but I think they obscure rather than clarify any distinctions between the three.

situations are simply not suited for equilibrium analysis. Returning, then, to the two-stage game, while it may be evident
how to proceed for part of the game, the formal requirements
of a Nash equilibrium are that it is evident how to proceed for
the whole, and what is evident in part leads to a situation in
which there can be no evident way to continue. One is led to
ask, Is the requirement of Nash equilibrium too severe? Is there
something less restrictive that would characterize behaviour in
games in which there are 'self-evident part-strategies'?

(2) Chess, thought of as a game of complete and perfect information, is too complex a game for the players. Certain parts
of the game, e.g. various endgame positions, are simple enough
to be analysed in this fashion. But the game as a whole strains
the abilities of players to examine completely all the options
that they have.

(3) The variation on the cities game, played once without negotiation by two members of the audience of a lecture at Oxford, is not specially complex as to its rules. But it has many
equilibria, no one of which seems to be suggested by experience or deduction. If we allowed the two players to negotiate
before making their lists, equilibrium analysis (or, rather, the
prediction that they will achieve an equilibrium) would seem
much more likely to be borne out. If we kept selecting pairs of
players from the audience, we might imagine that some partition rule would take hold. Since these cities are all from Brazil,
if the two players were Brazilian geographers, or if they had
at their disposal a map of Brazil, then we might imagine that
they will achieve some focal-point partition. But played in the
circumstances described, there seems to be no evident way for
the players to act.

(4) Next is the game in which players bargain (in a single-
round, simultaneous-demand protocol) over 100 chips, each
representing a 1 per cent chance at a prize, with A's prize
being $40 and B's, $10. This is just like the variation on the
cities game; if players can't negotiate beforehand and they have

never played nor seen played this game, there seems no basis on which to select from the very rich set of Nash equilibria. If there is pre-play negotiation or if players have experience in this (or similar) situations, then bargaining might well resemble an equilibrium, as is clearly demonstrated by the Roth and Schoumaker experiments.

(5) If we think of the centipede game as a game of complete and perfect information, as depicted in Figure 4.12, then equilibrium analysis is inappropriate because the model being employed is a poor model of reality given the structure of the game. That is, theoretical analysis shows that the set of Nash equilibria of this game changes drastically with 'small changes' in the formulation such as, A assesses probability 0.0001 that B is the soul of co-operation. If players entertain such doubts about their opponents, and it does not seem unreasonable to suppose that they do, then we ought to include this uncertainty in our model.

If we do so, will equilibrium analysis be appropriate? This largely empirical question is addressed by McKelvey and Palfrey (1990), and Camerer and Weigelt (1987) present evidence on a related problem; evidence that so far doesn't lead to a strong conclusion one way or the other. But I am somewhat dubious in principle (unless we construct the model to be analysed *after* seeing the behaviour to be predicted). The range of possible assessments of A concerning B and of B concerning A is very wide, especially when one takes into account A's assessments of B's assessments of A, and so on. Is it reasonable to think that A and B will begin with conjectures that are consistent with those specified a priori in some model? And, if so, is it reasonable to think that A and B will be able to see through the maze of complex calculations required to find their own best course of action? Once we add small probabilities that A acts in this way or that, and similarly for B, equilibrium analysis becomes almost impossible. Moreover, with a model rich in possible 'behaviours' by A and B assessed by the other, it is not clear what equilibrium analysis will preclude, and so it is not clear that one could test the proposition.

This is not to say that equilibrium analysis of such games cannot suggest hypotheses to be tested. Indeed, there are some similar instances in which the predictions of Nash equilibrium analysis are fairly robust to the specification of player's assessments about each other, e.g. when one player A plays a sequence of opponents, each of whom plays only once. The analysis of this situation, performed by Fudenberg and Levine (1989), makes the strong prediction that A can act as a 'Stackelberg leader', as if A could precommit to whatever strategy she wishes. This is clearly sufficiently hard-edged to be testable.

I find it difficult to think through these issues in the context of the centipede game, because the game becomes forbiddingly complex if perturbed by the addition of slight uncertainties by the players about others. So let me put similar questions in a somewhat simpler context.[10]   Consider the two-player, simultaneous-move game depicted in Figure 6.2. Upon inspection, you will see that in this game the combinations U-W and D-Z are both Nash equilibria. It turns out that this game has a third Nash equilibrium, which involves both players playing random strategies; in this equilibrium, B randomizes between W and X.[11] The one strategy in this game that is not part of any Nash equilibrium is Y for player B.

Yet imagine staging this game as follows: From a given population—say, members of the audience of a talk at Oxford—pairs of individuals are chosen at random to play the game once and once only, without any information about how others have played this game and without any opportunity to confer. The payoffs in the cells are in units of pence. Based on casual empirical experience (with pairs chosen from the population of MBA students at Stanford University and payoffs in pennies),

[10] I risk overdoing this point. If you are already a bit glassy-eyed, move on to the discussion of deregulation of the US domestic air-transport industry.

[11] If you don't know what a randomized strategy is, don't panic. Just take my word for the next sentence in the text.

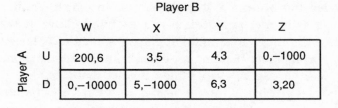

FIG. 6.2. A strategic form game

I would predict that in this setting players B will choose Y, the one strategy that is part of no Nash equilibrium, at least 80 per cent of the time.

The reason is simple. Strategy Y is optimal (assuming B tries to maximize his expected payoff) if the probability assessed by Y that A will choose U is anywhere between 0.017 and 0.998. The facts that there are multiple equilibria in this game and there is no evident way to select among them a priori imply that B has a very hard time predicting what A will do, and when B is even slightly uncertain, Y is best.

Of course, if A sees this, then she should choose D. Which then, one might argue, might lead B to Z as the safe choice. But in the casual staging of this game that I have observed, players A choose U often enough to justify the choice of Y by B. The question then becomes, Why does A choose U? Reasons can be supplied that make this example similar to the 'perturbed' centipede game. Suppose that, with small prior probability— one chance in fifty, say—A will choose U simply because she wishes to try for the payoff of 200. That is, the value to her of trying for this payoff outweighs the loss of 2p that she suffers if, as expected, B chooses Y. If we perturb this game to include this small chance in the mind of B a priori, then Y is indeed an equilibrium choice. The set of equilibria in this game shifts with 'small' changes in the formulation of the game (in terms of the possible utility function of A), in this case because of the enormous consequences to B of an 'error'.

Given this we can ask, Is this a case in which equilibrium analysis is in fact appropriate and useful, as long as we are intelligent enough to admit that players sometimes play games with payoffs other than money and we anticipate that some As may well choose U, even if they predict that B is very likely to choose Y? This is for the reader to judge; my opinion on this question continues to change with each passing day. To answer this question it would be helpful to ask of those As that choose U, 'Do you anticipate that B is very likely to choose Y?' That is, do such As choose U *despite* 'equilibrium conjectures' of how B will act, or do such As engage in somewhat wishful thinking? Unhappily, I have never asked this question of my students and so leave this entire issue open.

(6) We come finally to analysis of the US domestic air-transport industry following its deregulation. This is a case in which almost everything conspires against the fruitful application of equilibrium analysis. The players were forbidden by law from negotiating. The options that they had were extraordinarily complex and probably unclear to them. The objectives of their opponents were unclear, the evolution of the law was unclear, and so on. One can attempt to model all these uncertainties within a game-theoretic model, but (once again) it is probably only with the benefit of hindsight that we would be able to construct the 'right' model, where the model is deemed to be right if an equilibrium of it predicts how the airlines behaved. Since we have the benefit of hindsight in this case, you may be sceptical of my claim. But then (although this is an exercise with a finite life) predict the evolution of the inter-Europe air transport industry in the decade from 1990 to 2000.

The case for bounded rationality and retrospection

With all this as background, I can state the main thesis of this chapter. We must come to grips with the behaviour of individual agents who are boundedly rational and who learn from the past—who engage in retrospection—if we are to provide answers to questions like: When is equilibrium analysis appropriate? How do players select among equilibria? How do players behave when equilibrium analysis is inappropriate?

A reader of an early draft of this book (Ariel Rubinstein) challenged me at this point to provide a definition of the terms *bounded rationality* and *retrospection*. I mention the challenge because the task seems to me to be very difficult. The literature provides us with many definitions of bounded rationality, definitions which are not always consistent. I personally like Herbert Simon's dictum that boundedly rational behaviour is behaviour that is *intendly rational, but limitedly so*. That is, the individual strives consciously to achieve some goals, but does so in a way that reflects cognitive and computational limitations. Does this encompass all forms of *procedural behaviour*, in which the individual acts according to some procedure?[12] That is, is the line dividing bounded rationality and irrationality drawn somewhere between behaviour that follows any coherent procedure and behaviour that is simply chaotic? Or are some procedures (think of instinctive procedures) too far removed from rationality to be included within the term bounded rationality? I have nothing very useful to say here; I personally would choose not to include instinctive, procedural behaviour, but that is not a position with much thought behind it. Clearly, the type of behaviour that I will describe in the next section falls inside my own sense of boundedly rational behaviour, but I cannot offer a tight definition of what is in and what out. Nor do I have a tight definition of what constitutes *retrospective* behaviour; here I use the term loosely to mean behaviour where the past influences current decisions.

Notwithstanding my inability to provide precise definitions, it seems clear to me that the general (albeit vague) phenomena is crucial to the questions we have raised. The importance of coming to grips with bounded rationality is obvious from

---

[12] I intentionally forbear from using the term *procedural rationality* here.

the example of chess; we must take into account the cognitive (in)abilities of the participants in competitive situations if we are to separate out situations where the participants are able to evaluate all the options that they consider they have from those where they cannot. Indeed, the loaded phrase 'all the options they consider they have' is never more loaded than here—when we model a competitive situation by simplifying some options and ignoring others, we are using our instincts and intuition to guess how the participants themselves frame the situation they are in. Insofar as their frame is determined by their cognitive (in)abilities, we are taking their bounded rationality into account implicitly if not explicitly.

But there is more to the case than this. Stories that hold that players 'know what to do and what to expect' because of relatively direct experience are manifestly stories about how past experiences guide current expectations and hence current actions. At least to understand those stories, we must come to grips with the ways in which individuals learn from the past in order to predict the future. While one can pose 'hyperrational learning models' for such situations,[13] common sense suggests that learning from the past is a fantastically complex problem, often well beyond the cognitive powers of individuals to accomplish 'optimally'. The process of boundedly rational learning must therefore be confronted.

The importance of retrospection seems clearly related to stories about learning from direct experience. But what about the other stories why strategic expectations may be sufficiently aligned so that Nash equilibrium analysis is appropriate?

Whenever behaviour is governed by social convention, retrospection broadly defined is surely at work. Korean students know that they should defer to their professors, and their professors know that their students will defer to them; the experience of generations of students and professors (and other, more general, social hierarchies) so teaches both parties.

[13] See, e.g. Bray and Kreps (1987) and Crawford and Haller (1988).

Turning to focal points, the salience of a particular focal point in a given situation is clearly determined by the identities of the players, their cultural experiences, and so on. I don't know whether one can build a completely adequate theory of focal points based on a model of behaviour that is boundedly rational and retrospective (certainly I cannot), but I cannot see that such a theory will be satisfying without coming to grips with how one's past experiences and one's knowledge (however imperfect and inferential) of others' past experiences makes one focal point salient and others not. (I will discuss focal points further towards the end of this chapter.)

Concerning pre-play negotiation, I would contend that this explanation leaves open the very interesting question, What will be the outcome of negotiations? It is not clear that we will ever have anything much to say about this; however if we do, then we must step back to consider what drives the process of negotiations. As I tried to indicate in Chapter 5, and as the Roth and Schoumaker experiments clearly indicate (in a very simple setting), retrospection is probably a crucial element to bargaining. What a player believes is just or fair or his due and what is due his rivals in a bargaining situation is likely to be very important in determining the outcome of a particular negotiation. And such notions are apt to be determined by past experiences, precedents, and the like.

And as for deduction and prospection, I leave this to later, although let me make the (overly provocative) assertion that the application of deduction and prospection is an example *par excellence* of retrospective, boundedly rational thinking.

I don't wish to oversell this case (and fear I am in danger of doing so). All the problems of game theory are not going to be resolved by this tonic. Indeed, I have not yet discussed many of the problems raised in Chapter 5. We will not fully answer even the questions of when and how strategic expectations come into line by coming to grips with behaviour that is boundedly rational and retrospective. But it seems patently obvious to me that we will not fully answer these questions (and other important questions about game theory) unless we do so. On that basis, this seems an important frontier for the discipline.

This does *not* qualify as a brilliant insight. I am not the first economist to note the importance of boundedly rational behaviour to all manner of social and competitive interactions. I am not the hundred and first, and I am probably not the thousand and first. Others have beaten these drums and sounded these trumpets for many years. Indeed, the number of times this charge has been sounded is cause for pessimism; why should we expect progress on an issue that has been so clearly on the agenda of economists for so long?

There are some grounds for optimism. In many ways, these problems are right at the surface of non-cooperative game theory; we can hope that they arise in this context in a form simple and pristine enough so that we can make some progress. Moreover, there is much stirring in the recent and current literature about these issues, attacking these problems with a number of different paradigms. In the remainder of this chapter, I will sketch out one general line of attack that I find particularly attractive.

A line of attack

There have been a number of different research strategies recently adopted by economists and game theorists for capturing notions of bounded rationality. Some have used the metaphors of automata and Turing machines to model individual behaviour and to develop notions such as the complexity of a given strategy. (See, e.g., Abreu and Rubinstein (1988), Kalai and Stanford (1988), Neyman (1985), and Rubinstein (1986).) A different approach has been to enlist work on game theory from the literature of biology, stressing Maynard Smith's notion of evolutionary stable strategies in particular. (Friedman (1990) gives a good survey with extensions.) I will relate here a third approach.

The general scheme in this third approach places each individual player into a dynamic context. At any point in time, the individual is engaged in some short-run competitive interaction. In choosing his actions in the short run, the individual builds a model of his choice problem, a model which is typically a simplification or a misspecification (or both) of the 'true situation'. The individual finds his 'optimal' choice of action within the framework of this model and acts accordingly in the short run; the quote marks around the word 'optimal' should be noted, they will be explained in the next paragraph. In the longer run, the individual gathers information from his experiences and uses this information to 'improve' or update the model employed in the short run. The updating procedure is heuristic; some reasonable but not globally optimal method of adaptive learning is employed.

In some cases the individual is aware of the limitations of the model he employs and makes allowance for those limitations in some heuristic fashion. For example, in a model to be developed below, each individual learns from experience how others act, and his own choices in each period determine in part what information he might receive. A completely rational individual would make short-run choices taking into account the value (for the future) of the information so acquired. But, it is supposed, individuals may be unable to deal with the fantastically complex computations required to determine the value of information.[14] Hence short-run actions are largely guided by short-run considerations, but are shaded a bit to account heuristically for the unknown and incalculable value of information. (An example will be given below if this is opaque.)

---

[14] If you know the terminology, these computations will be more complex than solving a multi-armed bandit problem, because in this model, the 'multi-armed bandit' is one's rivals who are simultaneously solving similar problems about the first decision-maker and each other.

Note how this framework operationalizes Simon's dictum that boundedly rational behaviour is *intendedly rational, but only limitedly so*.[15] The use of the model in the short run is meant to capture the notion that individuals act purposefully, to achieve some objectives; the limits of rationality are captured by the facts that the model is imperfect, is adapted imperfectly, and may even have its results shaded heuristically.

It is perhaps evidence of the worst kind, but note that this is not a bad description of how some decision-makers (e.g. MBA students at Stanford University) are told to use models and analysis normatively; build a model, find the course of action that the model suggests and shade its recommendations to account for things known to be omitted from the model, and update the model and analysis as time passes and more information is received. These decision-makers are taught that in building a model, one must balance the tractability and cost of analysis of the model with the amount of detail put in, a trade-off that itself is not amenable to easy analysis; the balance must be struck according to intuition (i.e. heuristically).

Historical antecedents and recent examples

This type of model of individual behaviour has been employed in a number of applications in the distant past and more recent literature of economics. It is hardly something new. Let me cite a few examples.

Cournot dynamics

Cournot (1838) is perhaps the earliest reference to a model of this sort. Within a particular context, he proposes the following very simple sort of behaviour: Players choose best responses

---

[15] I do not mean to assert that this sort of specification encompasses all forms of behaviour consistent with Simon's dictum.

to the actions taken by their rivals in the period immediately preceding. In terms of the general scheme outlined, players' short-run models are that opponents will act at a given time as they acted just previously; and these models are updated completely in each period. For certain games, including many specifications of Cournot's game between two duopolists, this leads to convergence to a Nash equilibrium. But the extremely changeable nature of the models players employ can lead, for example, to cycles; in the rock, scissors, paper game, depending on the starting-point, play either cycles on the diagonal or among the six other strategy combinations.

Fictitious play

Brown (1951) suggested a model in which players play a strategic form game at a sequence of dates, $t = 1,2,\ldots$, where (making some allowance for starting the process) each player in each period chooses his strategy to be a best response to the empirical frequency of the strategies chosen by his opponent(s). This is the so-called method of *fictitious play*. Once again, the connection to the general scheme proposed is clear. In each period, players' models are that their opponents select actions according to the historical frequencies; players choose optimal responses to those models, and models are updated as more data become available.

As with Cournot dynamics, if play in such a dynamic setting settles down to a particular strategy combination, then that combination must be a Nash equilibrium. Indeed, if play bounces around but empirical frequencies settle down, then those frequencies must constitute a (mixed) Nash equilibrium. But Shapley (1964) provided a two-player, three-by-three strategic form game in which the method of fictitious play gives empirical frequencies that do not settle down at all, but 'cycle' instead (in quotes because the cycles take an ever-increasing amount of time to get through, as more and more previous history must be overcome).

General temporary equilibrium

One can think of the framework of general temporary equilibrium (Grandmont 1988) as being of this type. In each period spot and asset markets clear, where individuals value the assets according to models they have of returns on the assets and the value of wealth in the next period, models that are updated as time passes.

Bray's model of learning a rational expectations equilibrium

Another prominent set of examples concerns learning rational expectations equilibrium. (See Bray (1990) for a survey.) I give here a loose rendition of the model of Bray (1982).[16]

Imagine a sequence of two-period economies. In the first period of each economy, assets are bought and sold. These assets pay a return in the second period of the economy, which is then consumed. Imagine that there are two assets. One is riskless and pays a certain return of r. The second is risky; its return is (say) normally distributed with some given mean and variance. There are two types of consumer-investors in this economy. Some begin in the first period with no information except for what is given above; others begin with some private information about the return of the risky asset which changes from one economy in the sequence to the next.

Imagine that consumers, based on their own information, trade to equilibrium in the market for the riskless and risky asset in the first period. At any price for the risky asset (with the riskless asset as numeraire), those consumer-investors who are privately informed will condition their purchases on the extra information they have. If their information is relatively optimistic concerning the returns the risky asset will pay, they will purchase more of the risky asset. If their information is relatively pessimistic, they will purchase less. Hence for

---

[16] As always, my rendition is not accurate in all respects; the reader should consult Bray (1982) and also Marcet and Sargent (1989) for precise treatments.

markets to equilibrate, when the information held by the informed consumer-investors is relatively optimistic, the price of the risky asset will be relatively high; and when their information is relatively pessimistic, the price of the risky asset will be relatively low.

An uninformed consumer-investor living through the sequence of these economies will buy relatively little of the risky asset when its price is high and relatively more when its price is low. It is then not unreasonable to suppose that uninformed consumer-investors will begin to notice that the risky asset pays relatively more when they are holding less of it and that it pays relatively little when they hold more. Noting that there is a statistical dependence between the price of the risky asset and its return, these investors may begin to change their demand pattern; they will demand more than before when prices are high and less when prices are low, in anticipation of higher and lower returns, respectively. Of course, if there are many uninformed consumer-investors, their actions in this regard will change the equilibrium statistical relationship between the price and return of the risky asset.

A *rational expectations equilibrium* is said to hold when prices and returns of the risky asset somehow settle down to a stable statistical relationship; and where the uninformed consumer-investors correctly anticipate and act upon this statistical relationship. This notion of equilibrium has been used in a number of contexts in the literature (see, e.g., Admati (1989) or Grossman (1989)), to analyse a number of economic questions. A number of authors have asked the question, How is it that the uninformed come to learn the correct statistical relationship between prices and returns, especially when their learning (and use of that information) affects the relationship?

Bray (1982) investigated the following story. Suppose that at that start of economy t (for $t = 1,2,\ldots$ indexing the generations of this sequence of two-period economies), the uninformed consumers believe that the prices and return of the risky asset have been and will be related according to

$$r_k = a + bp_k + \epsilon_k,$$

where r_k is the return and p_k the price of the risky asset in economy k, ϵ_k is some random error term (normally distributed with mean zero and fixed variance and independent from period to period), and a and b are unknown constants. Note well, the uninformed consumer-investors suppose that this linear statistical relationship is time-invariant. Then *if* the uninformed consumer-investors have a good idea what are the values of a and b, given any price p_t for the risky asset in period t they could use this statistical relationship to formulate their demand. But how do they come up with estimates of a and b?

Bray supposes that the uninformed consumer-investors have access to the history of past prices and returns, $\{(\hat{p}_k, \hat{r}_k); k = 1, \cdots, t - 1\}$, where the hats denote realizations. Having survived an undergraduate course in statistical techniques in economics (and no more), the uninformed consumer-investors fit these data to the model $r_k = a + bp_k + \epsilon_k$ using ordinary least-squares (OLS) regression. This gives them estimates for a and b, which they then use in the short-run to formulate their short-run demand. Of course, this determines an equilibrium price p_t and when the period ends the risky asset returns some r_t; in economy $t + 1$ the uninformed act in similar fashion, updating their OLS estimates of a and b with these new data.

Note how this conforms to the general scheme outlined. At each point in time, the uninformed consumer-investors have a model of the world in which they live that is employed to make short-run decisions. Their model is that prices and returns in the current period (and the past) are related by $r_t = a + bp_t + \epsilon_t$ for given constants a and b. Moreover, those constants a and b are obtained by a heuristic procedure (OLS) from past experiences. The learning part of the story—the use of OLS regression—isn't quite 'right' in the full sense of the word.

In Bray's parameterization, in each period t there is in fact a linear relationship between r_t and p_t, but the coefficients a and b in that relationship depend in very complex fashion on time and the history of past returns and prices. When the uninformed consumer-speculators assume that these coefficients are time-invariant (and hence use OLS regression in very straightforward fashion), they are using a model of their environment that does not match the 'true' environment.[17] But what they are doing is not outrageous, especially if we think of them as not quite as smart about their environment as are we. (Perhaps they ended their training in economics in the early 1960s, before rational expectations equilibrium came into vogue.)

Armed with this model, Bray studied the questions, How will this economy evolve? Will the relationship between prices and returns settle down, and will the uninformed consumers learn this relationship asymptotically? (A nice feature of Bray's model is that if the relationship settles down, then the uninformed will learn it, since the 'true' relationship between prices and returns is linear in her parameterization, and OLS regression on a very large set of data will correctly recognize linear relationships. Compare this with the models of Blume and Easley (1982) in the same general context in which the price–return relationship can settle down and uninformed consumer-speculators *never* get it right.) She was able to show analytically that the price–return relationship settles down and the uninformed come to understand it if there are not too many uninformed. Otherwise the impact of their learning on the relationship between prices and returns causes that relationship to gyrate ever more wildly.

[17] I have put 'true' in quotes because the true environment depends on the model the uninformed are using; if we wish to assume that these uninformed consumer-speculators were using a correctly specified model of their environment, then we obtain a more complex problem in non-stationary rational expectations (see Bray and Kreps 1987).

Marimon, McGrattan, and Sargent's model of the medium of exchange

Yet another example is provided by Marimon, McGrattan, and Sargent (1989). They are interested in a fairly complex game connected to monetary theory. Consumers in a given economy are of several types; a consumer of a given type can produce one good and wishes to consume a good produced by another type. Consumers carry around a unit of one of the goods which they may consume; if they consume the good, they immediately produce one unit of the good that they are able to produce. There is a cost of carrying around a unit of each good, with costs varying with the type of good. Consumers meet at random and they may, by mutual consent, decide to swap the goods they are carrying.

Imagine a consumer who can produce and is carrying one good, say wheat, and who wishes to consume another, say (sweet)corn. Suppose this consumer meets a second consumer who is carrying rice and wishes to consume wheat. The decision of the first consumer whether to swap her wheat for the second consumer's rice will depend on whether she thinks that rice will be more valuable in the future for securing corn; is rice or wheat a better 'medium of exchange'?

Given a parametric specification of this economy, it is possible to find the Nash equilibria (in terms of consumption and trading behaviour), but it is not hard to imagine that the direct computation of such an equilibrium is beyond the wits of many consumers. Moreover, some parametric specifications will admit multiple equilibria. So Marimon, McGrattan, and Sargent propose a learning model for the consumers in this economy, to see if consumers following this model will learn equilibrium behaviour and, if so, which. The learning model is based on classifier models from the literature of pattern recognition. Each consumer carries around in his head a list of 'patterns'; each pattern is a description to some degree of precision of current circumstances and a recommended action. (For example, one pattern might be, If you are carrying wheat and you are offered rice, agree to an exchange. Another is, If you are carrying anything other than corn and you are offered rice, refuse to exchange.) At any point in time, to each of these patterns is associated a 'score'. Then as circumstances arise that call for a decision, the consumer compares the current situation with his list of patterns, seeing which match. (For example, a consumer carrying wheat who is offered rice would find that both patterns just described match.) When more than one pattern matches the present situation, the consumer chooses to carry out the action recommended by the matching pattern with the highest score; if no pattern matches, the consumer has a manner of inventing a new pattern which does fit (which is then applied), and this new pattern is added to the consumer's mental list and some low-scoring pattern previously on the list is deleted. Finally, the scores given to the various patterns are updated after each encounter according to how well they do; if following the recommendation of a pattern results in a good outcome (consumption of a desirable good), its score is raised, while if its recommendation leads to a bad outcome, its score is lowered.

This very rough (and somewhat simplified) sketch is probably too brief to give you a good sense of the model of behaviour used by Marimon, McGrattan, and Sargent, but you may be able to recognize in it the basic scheme. In the short run, the consumer's 'model' of the world is that there are certain actions that fit certain circumstances, and each action is graded by its score. In the longer run, those scores are changed heuristically according to the benefits and costs accruing to the various patterns of behaviour.

Marimon, McGrattan, and Sargent go on to apply this sort of model to some quite complex specifications of a trading economy. Their method of application is simulation. They set the economy off with some initial conditions, and then simulate random matchings and the resulting behaviour according to the rules of individual behaviour they specify, and they find, in general, convergence to a Nash equilibrium in consumption and trading behaviour of the economies they set up.

Fudenberg and Kreps' model of learning and experimentation in extensive form games

In work currently in progress, Drew Fudenberg and I have considered models of learning in extensive form games.[18] In the basic model, an extensive form game is given, such as the game in Figure 6.3. A large but finite population of individuals is imagined, some of whom take on the role of player A, others the role of B, and still others the role of C. We imagine a sequence of random matchings of the players, who in each matching proceed to play the game. Players begin with probability assessments over how their opponents will play, assessments that they update as they accumulate information through the course of play. We do not specify precise learning models, but instead restrict attention to models that satisfy the following general asymptotic property: As a player gains more and more data about how rivals play at some information set, the player's assessment about how his rivals will act at that information set asymptotically equals the empirical frequency of the actions he has observed. So, for example, if in the game in Figure 6.3 a particular player A has observed player B moving at his information set ten million times, and if B choose R' 6,039,433 of those times, then A's assessment of how Bs act in general is not far from an assessment that R' will be chosen with probability close to 0.604. This is like the model of fictitious play, except that we are looking at extensive form games and we only require that players 'believe' empirical frequencies asymptotically, if they have a lot of data.

Assume that players always choose short-run best responses given their current assessments of their rivals' actions. This

[18] This work is reported in Fudenberg and Kreps (1990). Obviously, everything related in this subsection should be attributed to that study.

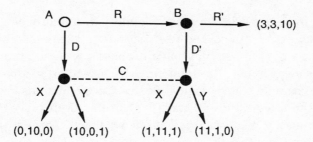

Fɪɢ. 6.3.   An extensive form game with interesting
properties for learning by direct experience

includes fictitious play as a special case, and Shapley's example establishes that convergence to some stationary behaviour or even limiting empirical frequencies cannot be guaran-
teed.   But because we are looking at extensive form games,
things are even worse than in the model of fictitious play; we
can converge to a situation that is not a Nash equilibrium.
Imagine that players A in the game of Figure 6.3 begin with
the hypothesis that Bs will probably play R' and Cs will probably play X. Suppose that Bs begin with the hypothesis that
As will probably play R and Cs will probably play Y. Then
any A's best response is R, and any B's best response is R'.
This reinforces the hypotheses of the As and Bs concerning
each other's actions, and it leads to no information whatsoever about what C will do. Hence the As' and Bs' assessments
about the actions of the Cs need never come into accord; they
can hold on to their disparate beliefs.   And (you can either
verify or take my word for it) this leads to stationary behaviour
(R-R') that is not a Nash equilibrium outcome of this game[19]
    The point is a simple one.   If we think that participants in

[19] In any Nash equilibrium, A and B must have identical conjectures about
how C will act, and then either A or B (or both) prefer that C will move instead
of winding up at the R-R' outcome. There is a mixed-strategy equilibrium that
gives R-R' with positive probability, but C must play with positive probability
in every Nash equilibrium.

some (repeated) situation share almost common beliefs about the actions of others because they share a common base of experience, then we may have problems with situations that do not recur relatively frequently. In particular, we may have problems with commonality of conjectures of actions taken 'off the beaten (equilibrium) track'. Hence assuming even rough commonality of conjectures about what will happen in such situations may be unreasonable unless we have some other story why conjectures will be roughly the same. Accordingly, in extensive form situations the formal notion of a Nash equilibrium may be too restrictive; we may wish to insist on commonality of conjectures along the beaten track but not off, which gives us a solution concept weaker than Nash equilibrium.

Alternatively, we can ask (and Fudenberg and I go on to ask) what it will take to get back Nash equilibrium in this setting. Roughly put, we require that players 'experiment', trying occasionally actions that are suboptimal (given current beliefs), in order to see what will happen. This means that players do not play precisely optimally given their short-run models; perhaps because they understand their cognitive limitations, they move a bit away from optimizing precisely in order to obtain information.

Given a basic story along these lines, we give conditions under which all 'stable points' are necessarily Nash equilibria and under which every Nash equilibrium is a possible stable point of some learning process of the general type we study. We go on to embellishments, exploring whether refinements of our basic story correspond to standard refinements of Nash equilibrium. (I will have something to say about our results later, but a capsule review is that some of the standard refinements do not fare particularly well.) We explore how players within a small group who play repeatedly might learn 'co-operative behaviour' in the sense of the folk theorem. And we give one further embellishment, to be discussed later.

Milgrom and Roberts's general analysis of learning

The Cournot model of dynamics, fictitious play, and Bray (1982) all posit very specific dynamic models of behaviour. Marimon, McGrattan, and Sargent (1989) are somewhat more general, but they are still relatively specific. One of the (self-described, mind) strengths of Fudenberg and Kreps (1990) is that the models of learning and, when it comes to it, of short-run behaviour are fairly general; we only specify asymptotic properties of learning, experimentation, and the short-run choice of action given a model of how others act, so that many specifications of behaviour would seem to be covered by our analysis.

Milgrom and Roberts (1988, 1990) go further along these lines. They study learning in strategic form games (although in Milgrom and Roberts (1990) they generalize their analysis to extensive form problems) and attempt to see what can be derived with minimal restrictions on the process of learning. For example, and somewhat roughly, they call 'adaptive' any learning process in a repeated play model with the property that for every integer n there is a larger integer n' such that strategies chosen after n' are best responses to strategies combinations by rivals that lie in the support of choices made by those rivals after time n. That is, any 'initial history' of play is eventually inconsequential in players' models of how others act. This encompasses both Cournot dynamics and fictitious play and (with some minor qualifications) all the models of learning and strategy choice considered by Fudenberg and Kreps. Of course, much more is included besides. And after posing this definition, they attempt to see what can be said about adaptive learning in general.[20]

With a definition that encompasses so many forms of 'learning', one might anticipate that very little can be said. But Milgrom and Roberts are able to give results of three sorts. First,

---

[20] The usual caveat applies; I am giving rough transcriptions of quite precise results, and you will have to consult their paper to see exactly what they do show.

they can show that adaptive learning (so defined) will eventually rule out strategies that are eliminated by iterated application of dominance. Second, they show how for a certain class of games the resulting behaviour can be 'bounded' by equilibrium actions. And third, when games of a related class admit a unique equilibrium (which happens in some very important classical games in economics) resulting behaviour must converge to the unique Nash equilibrium.

Objections to this general approach, with rebuttals

Objections to this general approach can be anticipated. Let me at least anticipate those for which I think rebuttals can be offered

Putting many objections into one sentence, analyses along these lines specify *ad hoc* behavioural dynamics and then either get very weak results or rely on simulations. Hence when all the work is done, either one doesn't know whether the 'results' are general or are only luck (if simulation is used) or one doesn't care about the 'results' (if weak results only are obtained). And in any case everything turns on behavioural dynamics that were dreamed up by the analyst, with no particular justification such as we have for the classic models of rational choice; such results as are obtained are thus very suspect.

Although you cannot tell without reading the papers cited in the last section, these examples were chosen in part to rebut this criticism. It is certainly the case that some analyses in this class rely on simulations and invoke *ad hoc* models of behaviour, but Bray (for example) gets quite strong analytical results,[21] Fudenberg and Kreps (1990) consider a fairly broad class of dynamic behaviour, and Milgrom and Roberts (1988, 1990) consider an extraordinarily broad class. The results in Fudenberg and Kreps are quite weak in some respects (after

---

[21] And Marcet and Sargent (1989) extend those results.

Shapley's example, there is no hope for a global convergence result), but Milgrom and Roberts show how very strong results can be obtained with fairly minimal assumptions for a specific class of games. That is, the criticism as stated is somewhat well taken, but I hope that if you look at the examples cited and others, you will see that the premisses of the criticism are not satisfied by all studies of this sort.

But I would venture to go a good step further than this. Even if a particular model is based on an *ad hoc* prescription of behaviour and is analysed with simulations only, this does not mean that one cannot learn a great deal from the exercise. Economic theorists (and I include myself) too often lose sight of the fact that while we may not be too restrictive in the behavioural assumptions we make[22]—and those typically made have an axiomatic base—we make enormously strong *ad hoc* assumptions in the range of actions we allow the actors in our models. The excuses we offer are often that we can use intuition guided by experience to factor in those things omitted from our model and, in the end, the test must be empirical. Why not employ the similar open-minded scepticism about simulations of *ad hoc* behaviour?

Similarities: Or, Deduction and prospection is bounded rationality and retrospection

A deeper criticism of this general approach is that it misses important parts of the puzzle. For example, what of games that admit an evident way to play owing to the powers of deduction and prospection?[23] Let me begin by repeating the

---

[22] *Pace* my more behavioural colleagues who are gagging at this statement; please put it down to a rhetorical device to get my less behavioural colleagues to come to a conclusion you will probably like.

[23] The ideas of this section are lifted from Fudenberg and Kreps (1990), and Fudenberg shares fully in any credit that attaches to it. (Since he doesn't have editorial control over this rendition, he is excused from any rhetorical excess.)

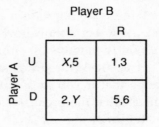

Fɪɢ. 6.4.   A 'game' with randomly distributed payoffs

earlier stated proposition that I mean to defend: Deduction and prospection as practised according to game theorists is bounded rationality and retrospection *par excellence*.

To make this point requires a bit of an excursion. Suppose the 'game' in Figure 6.4 was played repeatedly under the following conditions: Two members of the audience at an Oxford lecture play the game, with results made public. Then two more are chosen to play the game, and so on. The catch is that two of the payoffs in 6.4 are letters X and Y. Suppose that on each round of play, X and Y are chosen at random from the interval [0,10], each uniformly distributed and each independent of the other. Then there is zero probability that any game in the sequence of games played is precisely the same as a game played earlier. And so the method of fictitious play, or the sorts of models so far described as coming from Fudenberg and Kreps (1990) or Milgrom and Roberts (1989, 1990) do not apply.

It is not hard to see how we might change things to apply those models in this situation. Imagine that on the ten thousand and first round in the sequence, $X = 6.54$ and $Y = 3.4421$. If the current player A wanted to guess at how her rival B will act in this situation based on the ten thousand previous rounds of play, an intuitive thing to do is to look at previous instances of play when X was approximately 6.54 and Y was approximately 3.44. That is, we imagine that players suppose that games are 'similar' when they have approximately the

Player B

	L	C	R
U	0,0	10.1,10.1	0,0
M	10,10	0,0	0,0
D	0,0	0,0	10,10

(Player A = rows U, M, D)

(*a*)

Player B

	L	C	R
U	0,0	10,10	0,0
M	10.1,10.1	0,0	0,0
D	0,0	0,0	10,10

(Player A = rows U, M, D)

(*b*)

FIG. 6.5. Two 'similar' games with dissimilar evident ways to play

same payoffs and, when trying to predict what will happen in a given situation, players look at 'similar' situations.

This particular notion of similarity, based on the distance between payoffs, has some nice mathematical properties. (Some of these are developed in Fudenberg and Kreps (1990).) But one should not make more of this than is appropriate. For example, suppose you were set to play the game in Figure 6.5(*a*). A quite natural focal point is U-C. And in Figure 6.5(*b*), M-L seems 'natural'. Even though the payoffs in the two situations are quite similar, the qualities on which the focal points are based, unicity and Pareto-optimality of the payoffs, are discontinuous.[24]

The philosophical point is what is important here. If we rely solely on the story of directly relevant experience to justify attention to Nash equilibria, and if by directly relevant experience we mean only experience with precisely the same game in precisely the same situation, then this story will take us very little distance outside the laboratory. It seems unlikely

[24] I am grateful to David Canning and Hamid Sabourian for this observation.

that a given situation will recur precisely in all its aspects, and so this story should be thought of as applying to a sequence of 'similar' situations. Of course, to make sense of this, we must suppose that participants in competitive situations have some notion of which situations are similar and which are not, a notion that itself is built up from experience.

Indeed, when we pass to the justification that cites social conventions, we are imagining subjects who recognize situations as similar and then see how a given (presumably less than precise) social convention applies. Mistakes may sometimes be made, and in ambiguous situations players may actively seek to avoid mistakes.[25] But the general idea clearly involves judgement on the part of participants that (*a*) a particular general social convention does indeed apply in a specific setting and (*b*) it applies in a specific manner. Some sense of 'similarities' and/or 'suitability' is essential to this brew.

Taking a step even deeper into this particular morass, I contend that focal points are to some extent based on such calculations. Take the nine-cities game. Schelling (1960) argues that in cases such as this game, players are looking for some 'salient' rule for partitioning the list of cities. What makes a rule salient? Schelling suggests that the rule should suggest a clear division (how much more confident would you have been

---

[25] For example, imagine having pairs of players from various demographic groups play the game in Figure 6.6. Think of the payoffs as being in dollars. Players are not allowed to speak to each other, but each is given the curriculum vitae of the other and knows that the other is looking at the c.v. of the first. In this game, *if* players think they will have a common understanding of who should defer to whom, then we will see either M-L or U-C. But if they are worried that there is no clear rule to determine which of these is evident, then we may expect to see the Pareto-inferior D-R. Would a pair consisting of a Korean student and professor feel confident enough to try for the Pareto-optimal payoffs? For what other pairs is there sufficient clarity of some social convention? What if we picked a student out of a Stanford MBA classroom at random, allowed this student to record his or her selection (without revealing it), and then chose a second student? I am unaware of any experiments of this sort, and except for my cherished convictions about Korean students (admittedly based on only three cases), I have no particular conjectures to offer.

		Player B L	Player B C	Player B R
	U	−10,−10	5,10	−10,0
Player A	M	10,5	−10,−10	−10,0
	D	0,−10	0,−10	1,1

FIG. 6.6.   A game for measuring the strength of one's convictions regarding the applicability of a social convention of deference

in the Warsaw Pact–NATO rule if Berlin had been replaced by Budapest?); it should give a roughly equal division in a game of division (if Berlin was replaced by Brussels, Prague by Amsterdam, and Warsaw by Rome, would you have felt so confident with a division that gave Washington, Amsterdam, Bonn, Brussels, London, Paris, and Rome to one player and only Moscow and Budapest to the other?); most vaguely, it should be suggested by the presentation (pre-assigning Washington to one side and Moscow to the other makes this particular geographical split salient, while if Berlin had been assigned to one player and Washington to the other, one might have thought more about a split using alphabetical order or perhaps even one that depends on Axis vs. Allied capitals from the Second World War).

In all of these considerations, I would maintain that the players are, consciously or not, holding this very specific game up to some 'ideal division game' in which there is a single way to divide things equally. The players try out in their heads various rules for dividing the nine cities that make this specific game similar to that ideal division game, and when some rule makes the similarity fairly close, the player adopts that rule and expects his rival to do so as well.

Of course, this is hardly an explanation of the intuitive leap based on presentation; why geopolitical division and not the alphabet in this instance? But I think it has potential as an explanation for why certain division rules are rejected by the players; why Warsaw Pact–NATO would not be much in vogue if, for example, the list was Amsterdam, Bonn, Brussels, London, Moscow, Paris, Rome, Warsaw, and Washington.

Prospection and deduction is bounded rationality and retrospection

At last we come to prospection and deduction. I assert that when we, or the players, engage in deductive reasoning about a particular situation, we are using very complex 'similarities' that have been developed in the process of theorizing about games. Recall the discussion of incredible threats and promises in Chapter 4. We began with an intuitively obvious effect in one context that, when formalized, could be seen as well in other contexts and extended to more complex settings. That is, we looked (at a somewhat abstract level) for similarities which linked several contexts, so that an insight from one context could be imported to other apparently similar contexts. Think of such study as having normative instead of descriptive aims; that is, participants in some competitive situation study game theory in order to have a better idea how their rivals will act in a particular instance. Then this study is precisely the search for appropriately 'similar' situations.

But this study is deductive and not inductive, you may object. That is, we reason from first principles about what makes a particular threat less than credible. My response is that this is an incomplete description of how we reason. The formal basis for the similarity may be deductive, just as Euclidean distance between payoffs is 'deductive', but we do not rest there. We check the deductively derived similarity against past experiences, less formally by resorting to intuition or more formally by engaging in an empirical test of the theory.

Let me give two examples. Consider the game in Figure 5.4(*b*), reproduced here as Figure 6.7(*a*). The 'theory' of forward induction says that in this situation, player B if given the move will choose r, since B will reason that A must have chosen M in

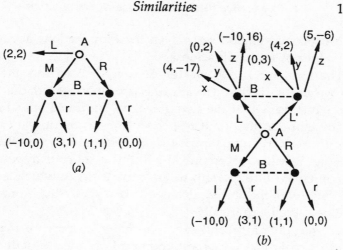

FIG. 6.7.   Two 'forward induction' games

an attempt to get a payoff of 3. In this case, we can come to a similar conclusion if we look at the game in strategic form and apply iterated dominance. (R is strictly dominated by L for A, and once R is eliminated, r weakly dominates l for B.) Do we believe in the validity of this prediction? That, clearly, is a matter to be settled empirically, but imagine that we looked at data which confirmed this prediction in this case. There is still the question, What is the *theory* that is at work here? Is it that iterated dominance works (at least, when the first round is strict dominance and there are at most two rounds further applied)? Or is it something more general?

There is a very sophisticated theory that predicts that B will choose r in the game of Figure 6.7(*a*), the theory of strategic stability (Kohlberg and Mertens 1986). It is completely beyond the scope of this book to explain this theory, but you can take my word for the following assertion: This theory applies equally well to the game in Figure 6.7(*b*); in this game, if A believes in this theory, A confidently predicts that B will choose r given the move in his bottom information set and B will randomize between x and y (each with probability one-half) at his top information set. Given these predictions about B's behaviour, A optimally chooses M.[26]

The question is, Do we (or does A) believe this theory? Do we (or A) accept as valid the principle behind this theory, which says that the two games in Figure 6.7 are similar in strategically important respects? In essence, do we (or A) accept this particular similarity? You are free to hypothesize whether it will work empirically in this case; your hypothesis and mine are unimportant to the current point, which is that the validity of this theory, this notion of a similarity, is one that will be settled empirically.

Let me give a second example that may obscure matters but that makes an important point on how we use deduction and prospection. Consider the game in Figure 6.1, which was cited previously as the canonical example of a game in which deductive thinking works. The principle applied is backward induction; B will choose R given the chance, and so A can safely pick D. But why do we (and, more to the point, A) know that B will choose R? We can reduce this statement to a tautology if we interpret the numbers in the figure as the players' 'payoffs', an ordinal expression consistent with each player's preferences as revealed by the player's choices. Then if B didn't choose R, B's payoffs wouldn't be as in the figure. But for practical purposes the application is made in a different fashion. Suppose we think of the numbers in the figure as dollar payoffs for the players. Then we (and A) deduce that B will choose R because we believe that B will prefer $0 to $−1.

---

[26] If you know about the theory of strategic stability, it applies in this case as follows. For simplicity, inflate the game tree by giving A an extra move at the start, where A chooses between the top and bottom halves of the tree. Then in the top subgame (so created) there is a unique Nash equilibrium, in which B randomizes between x and y with probability 1/2 apiece. This means that A's payoff in this subgame must be 2. Now apply equilibrium dominance or the never-a-weak-best-response criterion: In any equilibrium in which A chooses the top half of the tree with positive probability, choosing R is never a weak best response for A, hence R can be deleted from the game. And then B must respond in his bottom information set with r. The rest of the argument is trivial.

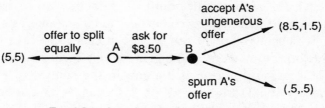

FIG. 6.8. A caricature bargaining game

Once we fall into this interpretation of how deduction is applied to predict the actions of players, we see how retrospection becomes crucial. Suppose the game was as in Figure 5.8(*a*), reproduced here as Figure 6.8. Interpret the numbers as dollar payoffs. Then the 'principle of backward induction' tells us (and A) to deduce that B will accept A's offer and so A should ask for $8.50. As we discussed in chapter 5, experience warns us against making this 'deduction' so cavalierly. If we think of the more complex, two-offer bargaining game discussed in Chapter 5, the literature (e.g. Ochs and Roth 1989) gives us a fair bit of evidence against the universal applicability of the theory that players always take more money instead of less.

You may object that in the question, Is this 'principle' empirically valid?, the principle at risk is *not* backward induction but rather the proposition that players always prefer more money to less. That is correct. But the application of deduction and prospection necessarily involves some hypothesis of the latter sort. So, if you prefer, the process of deduction and prospection *per se* is not an example *par excellence* of retrospective thinking. But the application of this process to make predictions about the behaviour of individuals must be.

The other problems of game theory

I have gone on too long, I fear, about the virtues of thinking about bounded rationality and retrospection when thinking about why a game might admit an evident way to behave (which would then legitimize Nash equilibrium analysis).

Or which might point the way to a concept weaker than Nash equilibrium. Both in the game which includes as a subgame the greatest-integer game and in the context of Figure 6.3, we saw that it might be useful to formalize the notion that pieces of a game might have a self-evident way to play, but that other pieces might not. Let us distinguish between the two cases: To deal with the game in Figure 6.3, we may want a solution concept reflecting the idea that pieces of the game that are rarely played—i.e. that are off the 'equilibrium' path of play—do not necessarily have an evident way to play: and thus various players' conjectures what will happen on those pieces might differ. For the game with the greatest-integer subgame, and more generally if we think of 'economic life' as a very complex game in which only some relatively isolable pieces have evident ways to play, we might want a solution concept that allows part of the solution to be evident and other parts not. I will not attempt a full discussion of these issues, but I mention all this here so that you do not prejudge the issue. The concept of a Nash equilibrium may well be too constraining, and any programme of justifying or explaining the Nash concept should be critical, willing to abandon the concept in cases where it asks for more than is reasonable.

Putting this question aside, what might the approach outlined here have to say about the other problems of game theory that were raised in Chapter 5? I don't have very much concrete to say, and most of what follows is wild speculation. (I can appreciate that you might feel that this doesn't differentiate what follows from what preceded by too much.) But if you are willing to settle for wild speculations, I have a few.

Disequilibrium behaviour

I have the least to say about modelling situations in which players have nothing like equilibrium expectations, and so I will dispose of this topic quickly. Presumably, periods of disequilibrium are those periods in which players begin with different expectations and are trying to learn what to expect of others. Hence it would seem clear that the behavioural specification of the learning process (and the disequilibrium choice of immediate actions) will strongly colour the flow of disequilibrium actions. And, therefore, researchers who dare to make predictions of such periods without resorting to some equilibrium model with hyperrational agents will be accused of ad hockery. This doesn't lessen the importance of understanding as best we can such situations (especially given what I will say in a bit about the inertial nature of equilibrium expectations and rules of the game), and while I have nothing concrete to say concerning such things, at least let me (along with many others) act the cheer-leader: For those who like high-risk research strategies,[27] there is much glory to be won in finding acceptable ways to move formal economics away from an insistence on complete equilibrium at all times.

The refinements

In Chapter 5, I developed the hypothesis that the problem with refinements (at least, in extensive form terms) is that we need some sense of why counter-equilibrium actions take place. If we are more explicit about the dynamic process by which players come to learn 'what to expect' and how they act as they are learning, then we will be developing a complete theory which will contain within it a theory of counter-equilibrium actions. That is, this approach to behaviour almost necessarily entails a complete theory.

For example, in the models in Fudenberg and Kreps (1990), counter-theoretical actions are manifestations of experiments by the players; experiments that are meant to gather information about how others play. Within the context of this model,

[27] I don't predict the expected return.

the refinement of subgame perfection arises fairly naturally.[28] And we are able to introduce further restrictions on learning that justify sequential equilibrium. But the forward induction refinements are quite dubious; if counter-equilibrium actions are the manifestations of experiments, and if one is more likely to undertake a particular experiment in a form that is thought will be less expensive, then something akin to Myerson's properness emerges.

I don't want (nor have I the space) to go into details. My point is simply that this line of attack for studying games carries with it a natural way to generate complete theories, which then promotes the study of refinements.

Equilibrium selection by other means

I can imagine at least two ways in which models of the type above can help inform us about the process of equilibrium selection by means other than refinements.

The first is obvious. Equilibrium selection will be retrospective in just the fashion seen in the Roth and Schoumaker experiment; the past experiences of the participants should inform how they will behave in the future. A sense of inertia in equilibrium selection enters the story that seems entirely intuitive.[29]

---

[28] Take this assertion with several large grains of salt. For technical reasons, Fudenberg and I restrict the class of games that we investigate to those in which each player plays at most once along each path through the tree. This restriction takes a lot of the steam out of the subgame perfection notion.

[29] Might this inertia be sufficient to overcome the dictates of optimizing behaviour that is part of the definition of equilibrium? A form of boundedly rational behaviour that is much discussed is *satisficing*, in which an individual settles for an action that gives a satisfactory outcome, even if a somewhat better outcome might be achieved. One story that rationalizes apparently satisficing behaviour with an optimizing story is that computation of a completely optimal course of action is costly, and the individual who finds a course of action that is sufficiently satisfactory is judging heuristically that the costs of further search (for something better) will exceed the benefits of the search. Another rationalization, suggested by models of retrospection and concomitant inertia in behaviour, is that the individual may settle for a 'satisfactory'

Second, I anticipate that the study of similarities and, especially, 'efficient' notions of similarity will help us to understand some of the process of equilibrium selection. Especially in an ongoing relationship, but in almost any social context, circumstances will change from one day to the next, and social conventions and *modi vivendi* are likely to emerge along lines that are better fitted to the range of circumstances, better fitted in the sense that they suggest clearly how parties are meant to proceed.

Let me illustrate what I have in mind with a very particular example. In the Stanford Graduate School of Business we have seven different fields of study for the Ph.D. In each of those fields, the behaviour of students and faculty towards each other is quite different; the responsibilities and expectations vary much more between fields than within.[30] For example, in the field of Accounting, there is a much tighter schedule about when thesis topics are to be found, when the job market is to be entered, and so on, than there is in the fields of Economics or Organizational Behavior. This may simply be the playing out of different equilibria in a situation that admits multiple equilibria. But my sense as an observer is that the relative homogeneity of the research process in Accounting, relative to that in the other fields, makes an equilibrium based on these expectations relatively more efficient.[31]

course of action because its consequences are well known and understood; to break from a routine might result in an increase in strategic uncertainty (how will others respond?), which is judged to be unworthwhile if the 'expected gains' are small. Notions of aspiration levels, arrived at by early experience and then rarely shifted, come into play. It goes almost without saying that the consequences of such considerations for how we view dynamic interactions could be enormous. If we look not for exact equilibria but only for almost- or epsilon-equilibria, the set of what is an equilibrium in certain contexts expands enormously. See, for example, the comments following concerning alternating-offer bargaining and changes in the bargaining protocol.

[30] I hear groans all around. How do I presume to measure these things? Where do my data come from? All this is impressionistic, although I hope it will accord with your intuitions about such things.

[31] I might claim that the hypothesis could be tested by looking for similar

If you know the literature on repeated play with noise, you may see an analogy to the general notion that reputation and similar constructions are most efficiently arranged when based on relatively less noisy observables (and hence are so arranged). At a less formal, mathematical level, if you are at all conversant with transaction-cost economics, you may recognize this as something of a variation on the tenet that transactions tend to arrange themselves in ways to accomodate on the net benefits of the transaction, net (in particular) of the transaction costs. Here I am saying something like, Equilibrium expectations (and hence equilibria) arrange themselves in a way that tends to economize on the costs of finding ourselves out of equilibrium because of mis-expectations. So, turning the analogy around, we come to the problem, Where do the rules of the game come from?

The rules of the game

Whenever a game-theoretic analysis of some situation is presented to the economics department seminar at Tel Aviv University, one of the faculty there, Abba Schwartz, can be counted on to object, 'If the players in this game are so smart, why are they playing this silly game? Why don't they change the rules and play a game where they can do better?' The question has been asked often enough so that it is part of the local ritual, and in fact the presenter is often let off the hook immediately with, 'Well, you can't be expected to answer this question.' But the question is a good one, and by now you will understand that I mean to deal with it by denying the premiss. People play 'silly' games because they are not quite so smart as we typically assume in our analyses. The rules of the game—who moves when, who proposes which contract, on which variables does competition focus, what are the

effects at other schools, except that the faculty at one school tends to come from others, so there is an enormous amount of 'cross-breeding of expectations' that makes different schools far from independent.

laws that provide the background for the transactions—these things are all akin to equilibrium expectations; the product of long-term experience by a society of boundedly rational and retrospective individuals. The tenet of transaction-cost economics—transactions *tend* to arrange themselves in a way that maximizes the net benefit of the transaction—is a guiding principle, where the inertia we see in institutions mirrors the inertia we see in equilibrium expectations, and the ways of groping for more efficient institutions—gradual evolution of institutions, the adaptation of institutions to sudden drastic changes in the environment, more conscious and purposeful breaking-out of well-worn equilibrium patterns and (perhaps) plunging into a period of disequilibrium, and everything between these—mirror similar sorts of changes to equilibrium expectations.

Ambiguous protocols

Finally, there is the problem that the theory requires precise rules of the game. As noted several times, we might hope for results saying that 'equilibrium outcomes' are not too sensitive to seemingly small changes in the rules of the game; results of this sort would give us the courage to make predictions what will happen in situations where the rules are *somewhat* ambiguous. Also as noted, the application of equilibrium analysis to alternating-move, bilateral bargaining seems to douse such hopes; seemingly small changes in the rules can give unintuitively large changes in the theoretically derived outcomes. In Chapter 5 I conjectured one way out of this seeming conundrum; specify some level of incomplete information about the two parties, and hope for results that show that with this sort of incomplete information, small changes in the rules don't matter.

Having climbed out on this limb, let me saw it off behind myself by suggesting that I will not be overly disturbed if my conjecture proves false, and in any case such results are apt to be more work than they are worth. Suppose we imagine individuals who do not optimize precisely but satisfice to some small degree. Suppose, moreover, that each individual's sense of what is satisfactory is driven by his experience in 'similar' situations in the past. Then it will be easy to show that small changes in the rules will be swamped by changes in the participants historically derived strategic expectations and sense of similarity. Indeed, my guess is that it will be too easy to derive such results; the real accomplishment will come in finding an interesting middle ground between hyperrational behaviour and too much dependence on *ad hoc* notions of similarity and strategic expectations. When and if such a middle ground is found, then we may have useful theories for dealing with situations in which the rules are somewhat ambiguous.

Final words

Non-cooperative game theory—the stuff of Chapters 3 and 4— has had a great run in economics over the past decade or two. It has brought a fairly flexible language to many issues, together with a collection of notions of 'similarity' that has allowed economists to move insights from one context to another and to probe the reach of those insights. But too often it, and in particular equilibrium analysis, gets taken too seriously at levels where its current behavioural assumptions are inappropriate. We (economic theorists and economists more broadly) need to keep a better sense of proportion about when and how to use it. And we (economic and game theorists) would do well to see what can be done about developing formally that sense of proportion.

It doesn't seem to me that we can develop that sense of proportion without going back to some of the behavioural assumptions we make and reconsidering how we model the actions of individuals in a complex and dynamic world. This means confronting some of the most stubbornly intractable problems of economic theory. But the confrontation has begun —indeed, it is gathering steam—and so while I think we can be satisfied with some of what has been achieved with these tools, it is appropriate to be happily dissatisfied overall; dissatisfied with our very primitive knowledge about some very important things and happy that progress is being made.

Bibliography

Abreu, D., and Rubinstein, A. (1988). 'The Structure of Nash Equilibrium in Repeated Games with Finite Automata'. *Econometrica*, 56:1259–82.

Admati, A. (1989). 'Information in Financial Markets: The Rational Expectations Approach'. In S. Bhattacharya and G. Constantinides (eds.), *Financial Markets and Incomplete Information: Frontiers of Modern Financial Theory*. (Totowa, N.J.: Roman & Littlefield), 139–52.

Aumann, R., and Shapley, L. (1976). 'Long Term Competition: A Game Theoretic Analysis'. Mimeo. Rand Institute.

Ausubel, L., and Deneckere, R. (1989). 'Reputation in Bargaining and Durable Goods Monopoly'. *Econometrica*, 57:511–32.

Bain, J. (1956). *Barriers to New Competition*. Cambridge, Mass.: Harvard University Press.

Banks, J., and Sobel, J. (1987). 'Equilibrium Selection in Signaling Games'. *Econometrica*, 55:647–62.

Blume, L., and Easley, D. (1982). 'Learning to Be Rational'. *Journal of Economic Theory*, 26:340–51.

Bray, M. (1982). 'Learning, Estimation, and the Stability of Rational Expectations'. *Journal of Economic Theory*, 26:318–39.

—— (1990). 'Rational Expectations, Information, and Asset Markets'. In F. Hahn (ed.), *The Economics of Missing Markets, Information, and Games* (Oxford: Oxford University Press), 243–77.

—— and Kreps, D. (1987). 'Rational Learning and Rational Expectations'. In G. Feiwel (ed.), *Arrow and the Ascent of Modern Economic Theory* (New York: New York University Press), 597–625.

Brown, G. (1951). 'Iterative Solution of Games by Fictitious Play'. In *Activity Analysis of Production and Allocation*, (New York: John Wiley & Sons).

Camerer, C., and Weigelt, K. (1988). 'Experimental Tests of a Sequential Equilibrium Reputation Model'. *Econometrica*, 56:1–36.

Cho, I.-K., and Kreps, D. (1987). 'Signaling Games and Stable Equilibria'. *Quarterly Journal of Economics*, 102:179–221.

Cournot, A. (1838). *Recherches sur les principes mathématiques de la théorie des richesses*. Translated into English by N. Bacon as *Researches in the Mathematical Principles of the Theory of Wealth*. London: Haffner, 1960.

Crawford, V., and Haller, H. (1988). 'Learning How to Cooperate: Optimal Play in Repeated Coordination Games'. Mimeo. University of California at San Diego. Forthcoming in *Econometrica*.

Friedman, D. (1990). 'Evolutionary Games in Economics'. Mi-meo. University of California at Santa Barbra. Forthcoming in *Econometrica*.

Friedman, J. (1971). 'A Noncooperative Equilibrium for Supergames'. *Review of Economic Studies*, 28:1–12.

—— (1977). *Oligopoly and the Theory of Games*. Amsterdam: North-Holland.

Fudenberg, D., and Kreps, D. (1990). 'A Theory of Learning, Experimentation, and Equilibrium in Games'. Mimeo. Stanford University Graduate School of Business.

—— and Levine, D. (1989). 'Reputation and Equilibrium Selection in Games with a Patient Player'. *Econometrica*, 57:759–78.

—— Kreps, D., and Levine, D. (1988). 'On the Robustness of Equilibrium Refinements'. *Journal of Economic Theory*, 44:354–80.

Gale, D. (1986). 'Bargaining and Competition, Parts I and II'. *Econometrica*, 54:785–818.

Grandmont, J.-M. (ed.) (1988). *Temporary Equilibrium*. Boston, Mass.: Academic Press.

Grossman, S. (1989). *The Informational Role of Prices*. Cambridge, Mass.: MIT Press.

Gul, F., Sonnenschein, H., and Wilson, R. (1986). 'Foundations of Dynamic Monopoly and the Coase Conjecture'. *Journal of Economic Theory*, 39:155–90.

Harsanyi, J. (1967–8). 'Games with Incomplete Information Played by Bayesian Players'. *Management Science*, 14:159–82, 320–34, 486–502.

—— (1973). 'Games with Randomly Disturbed Payoffs: A New Rationale for Mixed-Strategy Equilibrium Points'. *International Journal of Game Theory*, 2:1–23.

—— and Selten, R. (1988). *A General Theory of Equilibrium Selection in Games*. Cambridge, Mass.: MIT Press.

Hellwig, M. (1986). 'Some Recent Developments in the Theory of Competition in Markets with Adverse Selection'. Mimeo. University of Bonn.

Kalai, E., and Stanford, W. (1988). 'Finite Rationality and Interpersonal Complexity in Repeated Games'. *Econometrica*, 56:397–410.

Klemperer, P., and Meyer, M. (1989). 'Supply Function Equi- libria in Oligopoly under Uncertainty'. *Econometrica*, 57: 1243–78.

Kohlberg, E., and Mertens, J.-F. (1986). 'On the Strategic Stability of Equilibria'. *Econometrica*, 54:1003–38.

Kreps, D. (1990). *A Course in Microeconomic Theory*. Princeton, NJ: Princeton University Press.

Ledyard, J. (1986). 'The Scope of the Hypothesis of Bayesian Equilibrium'. *Journal of Economic Theory*, 39:59–82.

McAfee, P., and McMillan, J. (1987). 'Auctions and Bidding'. *Journal of Economic Literature*, 25:699–738.

McKelvey, R., and Palfrey, T. (1990). 'An Experimental Study of the Centipede Game'. Mimeo. California Institute of Technology.

Marcet, A., and Sargent, T. (1989). 'Convergence of Least Squares Learning Mechanisms in Self-Referential Linear Stochastic Models''. *Journal of Economic Theory*, 48:337–68.

Marimon, R., McGrattan, E., and Sargent, T. (1989). 'Money as a Medium of Exchange in an Economy with Artificially Intelligent Agents'. Mimeo. The Santa Fe Institute.

Milgrom, P. (1989). 'Auctions and Bidding: A Primer'. *Journal of Economic Perspectives*, 3:3–22.

—— and Roberts, J. (1988). 'Rationalizability, Learning, and Equilibrium in Games with Strategic Complementarities'. Mimeo. Stanford University Graduate School of Business, forthcoming in *Econometrica*.

—— —— (1990). 'Adaptive and Sophisticated Learning in Repeated Normal Form Games'. Mimeo. Stanford University Graduate School of Business.

Nash, J. (1950). 'The Bargaining Problem'. *Econometrica*, 18:155–62.

—— (1953). 'Two Person Cooperative Games'. *Econometrica*, 21:128–40.

Newhouse, J. (1982). *The Sporty Game*. New York: Knopf.

Neyman, A. (1985). 'Bounded Complexity Justifies Cooperation in the Finitely Repeated Prisoners' Dilemma'. *Economic Letters*, 19:227–9.

Ochs, J., and Roth, A. (1989). 'An Experimental Study of Sequential Bargaining'. *American Economic Review*, 79:355–84.

Osborne, M., and Rubinstein, A. (1990). *Bargaining and Markets*. Boston, Mass.: Academic Press.

Porter, M. (1983). *Cases in Competitive Strategy*. New York: Free Press.

Rasmussen, E. (1989). *Games and Information: An Introduction to Game Theory*. New York: Basil Blackwell.

Roth, A. (1979). *Axiomatic Models of Bargaining*. Berlin: Springer-Verlag.

—— and Schoumaker, F. (1983). 'Expectations and Reputations in Bargaining: An Experimental Study'. *American Economic Review*, 73:362–72.

Rothschild, M., and Stiglitz, J. (1976). 'Equilibrium in Competitive Insurance Markets: An Essay on the Economics of Imperfect Information'. *Quarterly Journal of Economics*, 90:629–50.

Rubinstein, A. (1982). 'Perfect Equilibria in a Bargaining Model'. *Econometrica*, 50:97–110.

—— (1986). 'Finite Automata Play the Repeated Prisoner's Dilemma'. *Journal of Economic Theory*, 39:83–96.

—— (1989). 'Competitive Equilibrium in a Market with Decentralized Trade and Strategic Behavior: An Introduction'. In G. Feiwel (ed.), *The Economics of Imperfect Competition and Employment: Joan Robinson and Beyond* (London: Macmillan Press, Ltd), 243–59.

Schelling, T. (1960). *The Strategy of Conflict*. Cambridge, Mass.: Harvard University Press.

Selten, R. (1965). 'Spieltheoretische Behandlung eines Oligopolmodells mit Nachfrägetragheit'. *Zeitschrift für die gesamte Staatswissenschaft*, 12:301–24.

—— (1975). 'Reexamination of the Perfectness Concept for Equilibrium Points in Extensive Games'. *International Journal of Game Theory*, 4:25–55.

—— and Stoecker, R. (1986). 'End Behavior in Sequences of Finite Prisoner's Dilemma Supergames'. *Journal of Economic Behavior and Organization*, 7:47–70.

Shapley, L. (1964). 'Some Topics in Two-Person Games'. *Advances in Game Theory, Annals of Mathematical Studies*, 5:1–28.

Spence, M. (1974). *Market Signaling*. Cambridge, Mass.: Harvard University Press.

Stahl, I. (1972). *Bargaining Theory*. Stockholm: Economic Research Institute.

Stiglitz, J., and Weiss, A. (1991). 'Sorting Out the Differences between Screening and Signalling Models'. In M. Bacharach *et al.* (eds.), *Oxford Essays in Mathematical Economics*. (Oxford: Oxford University Press).

Sultan, R. (1975). *Pricing in the Electrical Oligopoly*. Boston, Mass.: Harvard Graduate School of Business Administration.

Sylos-Labini, P. (1962). *Oligopoly and Technical Progress*. Cambridge, Mass.: Harvard University Press.

Tirole, J. (1988). *The Theory of Industrial Organization*. Cambridge, Mass.: MIT Press.

Von Stackelberg, H. (1939). *Marktform und Gleichgewicht*. Vienna: Julius Springer.

Wilson, R. (1977). 'A Bidding Model of Perfect Competition'. *Review of Economic Studies*, 44:511–18.

Zermelo, E. (1913). 'Über eine Anwendung der Mengenlehre auf die Theorie des Schachspiels'. *Proceedings, Fifth International Congress of Mathematicians*, 2:501–4.

Index